P9-EEM-038

SEASONS ~~~~ ~~~ OF JUPITER

a novel of India

SEASONS~~~~~
~~~~~OF JUPITER

by Anand Lall

GREENWOOD PRESS, PUBLISHERS
WESTPORT, CONNECTICUT

Originally published in 1958
by Harper & Row, Publishers, New York

Reprinted with the permission
of Harper & Row, Publishers

First Greenwood Reprinting 1971

Library of Congress Catalogue Card Number 78-109294

SBN 8371-3837-X

Printed in the United States of America

SEASONS ~~~~~ ~~~ OF JUPITER

Part I

THERE it was again — shrill, then mellowly warbling; a birdlike whistle from the garden across the narrow street. Still too weak to move about much, I craned forward in the large wicker chair and, above the chrysanthemums, I caught a glimpse of a very pleasant-faced elderly man with a twinkle in his eye, and ruddy friendly looking cheeks. Almost all the rest of his face was covered with a grey beard, and the long locks of hair on his bare head were thinning. The next moment, he was walking towards the house.

The house he was entering was dazzling white in the clear sunlight of a December midday in Delhi. I had noticed its new white stucco front as I was being driven home after having spent two months in the hospital recovering from a serious operation. Immediately my curiosity had been aroused and I had sat up straight, but my eyes involuntarily shut themselves against the sparkling whiteness. It was only as we were almost past the house and about to turn into our own gateway that I noticed that the garden, too, was blossoming under the new regime — tall sunflower plants and hollyhocks about to bud, rows of fullblown chrysanthemums, and a crowd of pink and white petunias like flocks of wide open butterflies.

Characteristically, my own people knew little about the house opposite. Whenever I inquired about it or drew attention to the new paint or gateway, or to the shiny black water spouts, there was simply a nod in reply,

9

followed by some such remarks as, 'Don't you think our roses are coming up beautifully this year?', or 'Would you like that new Kashmir counterpane for your bed?' It was as though the only impact the new house had made was to strengthen the pride and interest in our own home. I smiled to myself and gave up asking my mother any more questions. But a few days later, she made a chance remark that stimulated my curiosity.

'Strange man that Rai Gyan Chand in your new house opposite. Whistles with the birds! Queer.'

I decided that afternoon to walk for a while on our front lawn, from where I would see more of the house and garden opposite. I caught several glimpses of the elderly man, and twice I saw a young woman on their veranda. She had her black hair down — as though she had just washed it — and was dressed rather untidily in a loosely draped sari. I supposed she was Rai Gyan Chand's daughter. But all these inferences and glimpses added up to very little, and I was beginning to feel quite frustrated about the white house and its new occupants. I asked our gardener a direct question.

'Sita Ram, who has moved in opposite?'

'Young Maharaj, they say he is a rich man from Amritsar who owns two streets of houses and land and a factory. Probably tired of Amritsar and wants to live in the metropolis. Thinks he is a big man now!'

I nodded and went on with my stroll. It was pointless continuing the subject with Sita Ram who had imparted to me all he knew and had started to speculate on Rai Gyan Chand's motives. I found the afternoon's stroll invigorating and exciting. After two months in bed, how vast a space was that modest area of greenery and colourful flowers. To walk from the row of precarious chrysanthemum glory to the bony trunks and strong green leaves of the lime hedge was like a journey from one country to another, and there were many other areas to explore. Altogether, I felt much better, and decided to make my walk a daily practice.

On the succeeding days, I found myself looking over the hedges and across the street to the new white house. I saw new furniture arriving, and from time to time there were carloads of visitors — mostly countrymen, rather burly and pleasant-looking people. But most of all I kept my eyes open for Gyan Chand himself. He fascinated me as he walked through his garden or called out instructions to the servants or put his arm around the untidily dressed young woman who frequently had her hair down. Sometimes he would stand under a tree and whistle to the birds. On the third day of my garden peregrinations, I became conscious that, in return, Rai Gyan Chand was observing me.

At first, this knowledge made me uncomfortable. I tried to avert my gaze from the direction of the white house, but almost immediately my first reaction was succeeded by an entirely different feeling. I was pleased that he was looking at me. It was the beginning of communication, and surely that was what I really wanted. So we both looked at each other and to the boldness of our mutual staring were soon added involuntary smiles.

About a week later, when I was walking near our gateway, Rai Gyan Chand appeared just inside his own gate and called to me in a full, slightly lilting voice.

'I hope you are feeling much better now.' He had obviously made some inquiries and had learned of my illness.

Though I had been hoping for something like this to happen, I was somewhat overcome now that it had actually occurred. I stammered a reply. 'Yes. Yes, thank you, Rai Sahib. I am all right ... I mean I soon hope to be.'

The full voice resumed with a slower lilt. 'Of course you will be fully recovered soon. A young man like you will be strong and vigorous in no time. And these beautifully sunny days are just what you need.'

He was obviously ready to talk to me, and though he stopped at that point he looked as if he was searching my face deeply — keeping up the communication in that way. He nodded as though my looks had revealed something.

'And, of course, you need company, too. Well, why not come and visit us?' Then he smiled and added, 'Not that I, with my grey hair, could be much company for a young man, but do come if you would like to. How about tomorrow — at about three in the afternoon when the sun still has some warmth to send us — hm-m?'

'It is very kind of you, Rai Sahīb. I would like very much to come. I will call at three tomorrow,' I replied.

He waved to me and I raised my hands together in a *namasté*.

The next day, at three, Rai Gyan Chand met me in the tiled porch of his house, and together we went up the few low broad marble steps to the veranda.

'Come,' he said. 'We will go to the back garden and sit there.' He led me through the house.

If the front garden had showed signs of careful horti-culture, the back garden was where real diligence had been expended. The lawns were sparkingly green and trim; there were flowering shrubs, rare plants, and prize poppies and pansies in bloom. Facing south was a small marquee. Its south and east fronts were open — the west and north sides being shielded from the wind by tent walling. Under the marquee were three chairs, a low table, a small work table and some footstools — all on a rough woollen carpet.

'This is where I spend most of the day,' and he waved his hand to the work table and let it rise a little to indicate the trees and shrubs in the garden. My eyes followed his gesture and I noticed small wooden nesting places in some of the trees and shrubs.

We sat under the marquee. On the work table lay an open Sanskrit text and a manuscript in which someone had been writing in large, beautifully formed Sanskrit calligraphy. Noticing my inquiring look, Gyan Chand talked as if he were answering a question I had already formulated.

'Oh, that is my attempt at becoming a philosopher — it is my latest attempt.' And he smiled blandly, as though the position was now absolutely clear.

Baffled by his reply, and striving to assume a very mature expression meant to bridge the gulf between Gyan Chand's sixty years or thereabouts, and my own thirty, I said, 'I suppose we are all trying in our own way to be philosophers. But I must confess also that I am curious to know what it is ...' my last few words fell away from the mature tone that I had tried to assume.

'I will tell you,' he responded in his kindly and somewhat casual way. It was as if he had understood that if he were to react either enthusiastically or coldly to my too obvious curiosity, it would embarrass me painfully by exposing the breakthrough of my youthfulness.

He went on, in a slightly offhand manner. 'Many years ago, a friend of mine in Paris said to me, "If you want to be a writer, take one of Balzac's novels and copy all of it, word for word. You will then imbibe something of his art and technique." It occurred to me that I might transfer my friend's recipe to the field of philosophy. I have taken Samkara's *Crest Jewel of Wisdom* — there you see it lying open on the table — and I am faithfully copying it in the open manuscript book.'

To me it seemed a most ingenious idea. Indeed, as I had never really thought of setting out to be a philosopher,

13

any project of study in the subject was likely to be completely novel to me.

I burst out, 'Rai Sahib, what a wonderful and scholarly way of setting out on the road to philosophy.'

Gyan Chand laughed gently at my youthful extravagance. 'Ah, my friend. Not the road to philosophy. That road I have trodden for countless years. But it has been often like walking in a maze. It is the systematic insight of the true philosopher that I am after now, so I am copying Samkara's masterpiece. I work slowly and deliberately, never copying more than a page and sometimes doing no more than five lines in a whole day. So, you see, after two years or so I might become a bit of a philosopher!'

Then the conversation shifted and we talked of his early days in Europe. He fascinated me with tales about his friends, his professors, and European manners and customs. At about five in the afternoon, the young woman — wearing a loosely draped sari — appeared briefly on the back veranda. Gyan Chand saw her and his face brightened. I made my excuses and got up to leave. He did not ask me to stay, but getting up with a youthful alacrity surprising for his age he said:

'I am glad you came. I hope it might also help you to recover fully. You know how they say that pleasant conversation stimulates one's healthful glandular secretions! Come when you want to.'

Two days later as I was strolling near our gate in the early part of the afternoon, Rai Gyan Chand again hailed me. 'How are you today?' Seeing me smile and raise my hands to greet him, he went on, 'Good. Good. Then why not come over and let us continue our conversation?' His whole face looked as though it participated warmly in this kindly invitation.

I thanked him, and as before we went through the

house to the marquee on the back lawn. Again it was a warm sunny day and my blood seemed to respond to it as it had not done since I fell ill. I felt it almost twanging and singing as it coursed along. I craved for talk of life and adventure, and Gyan Chand had already said enough to raise more than a mild expectation of rich stories from his past. Listening to him I hoped, vicariously, to plunge into a wide and strange world.

I was not far wrong. It became my practice for the next several months, till I was fully recovered and ready to leave to enter a law firm in Bombay, to spend the afternoon with my friend Rai Gyan Chand. We grew very close to each other. I was frankly fascinated by his experiences and even more by the fact that at the age of about sixty he could look back on his past life without regret or longing. He accepted it all as part of himself and was still capable of adding to it: witness the new white house, the garden, and above all the young woman.

I got to know her only slightly. He referred to her as his wife, Devi. She seemed very reserved, but I soon learned that she was approaching the time of her confinement, which explained her loosely draped sari and her seeming carelessness about her own appearance. She was young, about my own age I thought, and very handsome in spite of being ungroomed. She seemed to live under Gyan Chand's spell. He charmed her and, at any rate while I knew them, nothing could have broken through the spell he cast over her.

Gyan Chand took to me in a way which it would have been impossible to anticipate. He had many friends and acquaintances, some he told me about, and many came and saw him from time to time. One or two special friends he visited regularly. But during that period he gave me more of his time than to all the others together.

Perhaps the most singular and moving incident in those

15

few months occurred one day when he had come to see me. He did so rarely, for his back garden had become established as our meeting place. We were walking on our own lawn. At first we talked about his philosophical studies, and there were casual remarks about the flowers and the song of a small white and yellow bird that sat on the slender green branch of a shrub, merrily pecking at the overblown flowers. Then Gyan Chand said to me:

'In the course of my life, one of the special studies I made was about sex.' He said this as if there were nothing special about it, but the slight tremor in his voice surprised me somewhat, for he had told me of his own life without reservation.

I looked at him inquiringly. Then he went on. 'In fact, a manuscript of mine on the subject is in the museum at X — at least that is what they tell me. I gave it to the Maharaja there to look at, but he died shortly afterwards and his son kept it. It was he who told me that it had been put in the museum. It could not be returned to me as there was no record of its having been merely on loan to his father. Under their laws it was his father's property.'

'Good heavens!' I broke in, 'and you have no other copy?'

'I haven't. But that does not matter to me. I have no need of the manuscript. I was thinking of it in connection with you. Perhaps you are hoping soon to marry. In olden times, sex was part of the course of studies that the would-be householder pursued. I do not know whether you have complied with the old injunction in this matter. But you will not mind my telling you that you do have a responsibility to the woman who will spend her life with you. I have been wondering whether you are equipped to play the role of her lover.' He stopped for a moment.

I was fairly free of prudery, but I did nevertheless feel a little uncomfortable at this last remark. What was I

16

expected to do? Tell him all I knew and also whether I was in robust health and vigour? I did not know, so I said so.

'I do not know, Rai Sahib.' Then to draw him out, I added, 'I did not know one required any special knowledge.'

Gyan Chand put his hand on my shoulder for a moment, and then sliding it down he squeezed my forearm for a moment before releasing it and reopening the conversation.

'My dear boy, so you haven't completed your studies. It might have been of some use to you if I could have given you my book.' Again he stopped, and this time it was obvious that he had something on his mind about which he was diffident. We walked across the lawn in silence and neared a lofty Arjuna tree which was being stormed by a homing flock of birds. And, as always, in such a situation there was a feathered babel above us. Gyan Chand looked up and smiled.

'Jostling and squawking for their mates — happily, mind you!'

Then he came to the subject on his mind. 'I do have a feeling of concern for you. You will go to Bombay and you will be lonely. Soon you will think you are in love, or some of your friends will find you someone they think you should marry. I don't want you to be plunged into a frustrated, unhappy marriage. I know you can take care of the material and intellectual needs of your wife. But there remains that which is engaging the birds in the Arjuna tree, and actually it engages our minds at least as much. After pondering deeply, what I have done for you is to sketch twenty-eight of the most pleasurable postures of union. In them, and remembering that you must not give way to impatience, you will not fail your wife.'

It was an amazing announcement. With a quick movement he took from his pocket a large sealed envelope.

17

'Keep it,' he said, 'and look at the contents later.'

I was too abashed and surprised to say anything. I stammered my thanks and hastily put away the envelope into my pocket — it was one of those ambiguous gestures that seeks to convey unconcern but really indicates secret eagerness. I wanted to rush to open the envelope and hoped that Gyan Chand would not prolong his visit. He must have sensed my excitement, for he put his arm around me affectionately and said in his gently lilting voice:

'I will go now to Samkara. I feel I want to copy a little more of the *Crest Jewel of Wisdom*.'

His hand released me and he walked quickly through the gate. As he passed under a tree in his garden, I heard a shrill chirruping from the bird-filled branches.

I immediately went to my room and opened the sealed envelope. Each on its separate sheet of paper were twenty-eight discreet but abundantly clear and power-fully executed drawings. They reminded me of the masterly erotic sculptures of some of our medieval temples, for they too had a symbolized quality in their portrayal of carnal pleasure. Some of the postures themselves were acrobatic and almost amusing; but what amazed me was the free, vigorous line of Gyan Chand's drawings. The old man was an artist: a talent I had never suspected in his talks about life, his search for beauty and philosophy. I had come to respect and feel much affection for Gyan Chand, and now, looking at these masterly studies of 'the artist in his bedroom', I spontaneously gave him — or at any rate his talent — my admiration. I put the drawings away carefully and they have always remained among my few treasured possessions.

The incident of the drawings occurred about three months after my return from the hospital, and when three weeks remained to the date of my departure for Bombay.

The next day I visited Gyan Chand at the usual hour, and we talked, strolled in the garden, drank tea, and then ate our evening meal together, but neither he nor I mentioned the contents of the envelope. I wanted to express my admiration for his drawings, but that would have meant referring back to our conversation of the previous day and I felt that the look in his eyes — a sort of clarity unambiguously given to the moment and to what we were talking about — shut out the events of the previous day. Nor was the subject mentioned during the rest of my stay at Delhi. I did ask him one day if he had done any painting or drawing, but he shook his head and nothing more was said about the matter.

It was in those last three weeks that the idea of writing down the story of Gyan Chand took possession of me. At first I paid no serious attention to it, but as the fullness of the story developed, the idea took stronger hold. My only hesitation was that if I were to try and tell it on his behalf, it would lose much of its flavour and subtlety.

After my departure from Delhi, I still hesitated for some months till it occurred to me that I should let him tell his own story — let him, Gyan Chand, do the narration in his own words; that the 'I' who has so far written on these pages should withdraw, giving place to the 'I' of Gyan Chand.

In the following pages, then, is recorded Gyan Chand's story in his own words, and my hope is that in this way most of it will be preserved accurately.

Part II

ALL my early life, that is to say till my father died when I was twenty-one years old, was largely conditioned by one chance incident — so much so that, in a sense, I became a victim of the power it exercised in our household. I must have been about seven years of age at the time.

Late one afternoon, after our pony ride in the park my elder brother, my sister, and I, were with my mother in her sitting-room on the second floor of our house in Amritsar. Suddenly there came a burst of shouting from our large front courtyard — it was a sort of private square — intermingled with the shrill trumpeting of Moti our elephant.

'Look out! Look out!'

'He will be killed!'

'Run for your lives!'

Instinctively, my mother put her arms around us. My brother and sister were petrified, but I wriggled out of my mother's grasp and dashed out of the room. She cried out, 'Gyan! Gyan! Come back!' but my excitement overwhelmed the urgency of her plea, and in a moment I was at the bottom of the stairs and rushing into the courtyard. Everyone else was running to its corners or making for the gate.

In the centre of the courtyard was Moti the elephant, swinging Janak, her keeper, in her trunk and blowing a soft frothy note of satisfaction. Everyone was certain that the helpless Janak was going to be dashed to the pavement, and then Moti would trample him and run amok, intoxicated with the scent of blood. Janak was a great hero of mine; and there he was swaying helplessly in the elephant's trunk while the victorious gleam brightened in Moti's eye. I was fired with righteous wrath. My whole body swelled with rage. My heart thumped like a big black hammer and the blood pounding through the veins in my neck seemed to pour riotously through my temples. I rushed almost directly under the elephant and shouted at him.

'Moti! Moti! Leave Janak! Leave him, I tell you!'

Moti, immensely surprised that one little person should come up and challenge him while all the others were in a panic of fear, put Janak gently down and walked meekly towards his stable as if to think over the meaning of his own and of my display. Someone rushed up and locked the stable door after him, and order was re-established.

Now the turbulent uncontrollable heaving of my chest and the choking of subsiding rage were acutely embarrassing. I was ashamed of myself and closed my eyes. But soon there was the sound of gentle admiring voices around me, and opening my eyes I found the servants had gathered about and were saying:

'Wah! Wah! Young Rai Sahib!'

'See how brave are even the little children of great people!'

'He's a young lion!'

My embarrassment overcame me, and darting through the admiring ring, I ran indoors to my mother. She was no longer anxious or angry. She, too, must have sensed what had happened, for she pressed me to herself happily.

That evening the servants, instead of exercising the special brand of authority wielded by old retainers, treated me with deference, almost with awe; and my parents and tutor glanced at me admiringly. I was puzzled at these new attitudes, for what I had done seemed to me perfectly natural and ordinary. But the rest of the household regarded the incident as a sign that I was endowed with unusual powers.

I soon ceased to be conscious of the event, but from time to time I realized that I was being treated in a very special way. Few young boys take kindly to school. My older brother objected strongly to being sent, but my father insisted. Normally, an unwilling younger brother would have been dealt with in a much more summary fashion. But when I in my turn kicked against going, there were long consultations between my parents and with our tutors. As a result it was decided, apparently on the ground that I had a unique instinct that enabled me to discern what I should or should not do, that I was to pursue my studies at home. My brother was deeply resentful of the favour shown to me, and from then on we grew rapidly apart. He consorted with his schoolmates, and left me to my own devices. I enjoyed learning languages — Hindi, Sanskrit, Punjabi, Persian, and some English — as well as history, geometry, and swordsmanship. In these subjects I read widely and probably knew far more than my contemporaries, but I was deficient in the sciences and in systematic philosophy.

All through my youth I was allowed to do more or less as I pleased, but at the age of eighteen came the crucial test of my special position in the household. The previous year, my elder brother — then approaching twenty — had been married to a bride of sixteen and sent off to manage one of the family estates near Ajnala, about twenty miles from Amritsar. Now my mother began alluding to my own forthcoming marriage.

'But Mother, I am not engaged. Where is the girl?' I said this to draw her out, but she took it to mean that I was keen to marry.

'Gyan, you will bring us much happiness by marrying and settling down to your share of family joy and responsibility. Do not worry about the girl. There are plenty of fine girls in our Punjab — you shall have the pick of them!'

I had no intention of marrying, but how was I to oppose my mother's feelings in this particular matter, certainly not directly: the marriages of her children were her prerogative. However, I had an idea that there was another way of handling this very delicate and portentous situation. I had read in the papers about young men from Bengal and even from Delhi who were setting out for England to undertake professional and other advanced studies. I decided that I, too, should travel abroad, perhaps acquire some professional training, and then establish myself in a position in which I could decide for myself when I should marry. The next day I told my father of my wish.

His peremptory face, formed by his habit of giving instructions and orders, flushed under his dark brow. 'But, my child, this is unheard of! No young man from Amritsar has been to England! Why should you go?' Under his annoyance I detected a current of apprehension.

I did not want to rouse his anger. I dropped my eyes and said gently, 'Respected Father, young men are setting out for England from Bengal and Delhi. Why should Amritsar be left behind the other cities of our country? I feel it would do us all some good if I were to go.'

My tactics had worked: 'Why do you feel that good could come of it, tell me, son?' asked my father with strange eagerness, searching my face as he questioned me.

I had had in mind no particular benefit to the family, having chosen a phrase merely because it sounded appealing. But seeing my father's eager look, it flashed across my mind that the words I had uttered were a

23

talisman. I recollected again the incident of the elephant, and I confess that, without any inner conviction or regard for truth, I exclaimed, 'Father, I feel sure that if I went abroad to educate myself great good would come to our family. I am quite certain of it.'

My father looked at me and nodded. He said no more and turned away, his face thoughtful and without a trace of displeasure. He was known to be an autocrat, and I am abashed now to think of how his workers, managers, and tenants stood before him in silence, never contradicting what he said.

For the next few days there were long discussions between my parents. Old family friends were called in and consulted. After a week or so quiet set in. I knew a decision had been taken, but an impassivity came over the faces of my parents. I could not tell whether it was because they had decided against letting me go or whether it was that they were strengthening themselves to bear the suffering that would be caused by the finality of an announcement that I could leave for England.

I had completed an hour's practice of swordsmanship and was re-entering the front veranda of the house. The exercise had relaxed me and the morning air was limpid and cool. I was musing to myself that the trip abroad was, after all, not worth a fight. It was just as well to marry and settle at home and take what life would bring. There was nothing my parents would not provide for me, and was not their affection invaluable to me? I would go now and tell my mother that if she still wished to seek out a beautiful, joyful and comforting wife for me, she would find in me a dutiful son. A flood of pleasure swept over me as I pictured her joyful surprise on hearing my announcement. I quickened my step and went up the stairs two at a time. Her sitting-room door was shut. I pushed it open as I called:

'Respected Mother, where are you!'

The door opened wide and my father stood before me, his face courageous and composed, but obviously suffering. Behind him was my mother, and for the first time I realized how the intimate sharing of experience can leave a single impress on two people: her expression was an exact image of his. My mood of enthusiasm was transfixed, and I was reduced to silence. It was my father who spoke in slow deep tones, with a slight lilt.

'Son — Gyan Chand. Out of our affection for you, we have decided that we should not obstruct your plan of embarking on a visit to England. We wish you well, and the blessings of God go with you. Only, we would beg of you, do not marry abroad, but hasten back to us as soon as your studies are completed.'

I was overwhelmed by this announcement. I glanced at my mother. The expression of courage on her face had gathered strength. I looked at my father. On his face, too, the expression had intensified to one of decision and tender affection. An inner flow of sadness poured through my heart. Why had I caused them to steel themselves to so difficult a decision? But as I looked at them again their strength seemed to staunch the bitter breaking flow within me. I tried to take possession of myself, but confused avowals of love, loyalty, and obedience swarmed through my mind. I could find no words.

My mother came up to me. 'Son, it will do good to all of us, as you say!'

I felt I had to speak now. 'Respected Father, and respected Mother, I will strive to make it good for us all. I love you both more dearly than my own life. Never would I think of marrying abroad and depriving my dear family of the right to settle me in life; and with all haste I will return to you!'

I emerged from the room into which ten minutes earlier

I had bounded filled with the thoughts of my decision to remain at home, but now my face was set irrevocably in the direction of a long voyage and a stay of many years in a strange land. The preparations advanced rapidly, and two months later I sailed for England.

I went determined to return quickly, and landed in England heavy with nostalgia. But once there, I found England congenial. The people were courteous and civilized. Only occasionally did I meet arrogant empire-conscious individuals who were of the opinion that I should occupy some lower stratum of existence. I realized that these were people who resented any intercourse distinguished by the grace and ease of which they felt themselves incapable. Besides, some of them had been in parts of the Empire in positions of greater or lesser authority in which their aloofness from the people was so complete that they had no idea of how an Indian, or the Chinese in Malaya, lived, much less how they felt. Of course, we Indians focused our attention solely on the arrogance and aloofness of the ruling imperialists, forgetting the contributory warp in this curtain of separateness for which we ourselves were responsible. For no respectable Indian family — and all families consider themselves very respectable — would have permitted the foreign ruler to be received as a friend and an intimate in the family. So, for many reasons, it was not unnatural that in England those who had some experience of India and other parts of the older world should resent the closeness of my relationship with English families.

To France, also, I paid many brief visits and was charmed by that country. But the agreeableness of France was very different from that of England. In England there was an easy quiet way, and almost effortlessly one could step out in rhythm. In France I had to use my wits every minute. It was exhilarating and amusing. I sensed that if I were ever to come a cropper in England I would be ignored and gently dropped, whereas if that were to happen in France I would be hounded and kicked by some and passionately taken up by others. Neither was the sort of experience that I particularly wanted, and, though there were ups and downs in my life, in both countries fortunately I was spared these extremes.

Having developed the ability to cope with the entirely different life of the West, it took hold of me like a strong habit. I did not realize what had happened till I found that, on one pretext or another, I was prolonging my stay abroad. I kept postponing my final law examinations at the Inns of Court. At home they must have sensed what had occurred. I began to receive cables asking me to return urgently. At first I sent polite rejoinders and pointed out that I still had examinations to take, but when it became high time for me to have finished my examinations I ignored the continuing cables and thereby confirmed my parents' suspicions that something was amiss.

Feeling there was nothing culpable about it, I enjoyed my pleasant idling, taking no pains to reassure my family. Besides, I shunned the effort it would require as being in itself an imposition upon the easeful indolence which I was practising.

This extraordinary state of affairs continued throughout the second year of my stay in England. But its end came suddenly. One evening, as I was setting out for a Sarah Bernhardt performance with a young woman of my acquaintance, a cable arrived.

'Oh, another cable. I won't bother to open it now,' I said as I flung it on my desk.

My companion showed surprise at this treatment of an urgent message, and she saw from the markings on the cover that it was a foreign telegram. She remonstrated with me.

Lightly I replied, 'You read it, then, and tell me what it says.' I expected it to be, as usual, plaintive, and yet as most of them were, reiterating their faith in my good sense.

She opened it. There was a moment of silence. Then she read in a strange voice, 'Your brother dead. Return at once. Father.'

After the theatre that night, I sent a long cable saying I would definitely take my last examinations within three months and then return. And truly I felt I could no longer delay my departure from England. I successfully passed the examinations, and in my state of elation it occurred to me that my brother had long since been dead and undoubtedly the severe blow to my parents had lost some of its edge with time; other arrangements had probably been made on the estate of which my brother previously held charge. In any event I knew my father would neither expect nor ask me, particularly after all these years abroad, to go back to run an estate located in the middle of the rural Punjab. All these 'arguments' appealed to me and I decided to spend a few last months enjoying myself in England and on the Continent. I proceeded to travel, and even started to write belles-lettres and poetry. It was a delicious experience, and I confess that the fact that I was being so atrocious to my parents added a savour of daring which excited me.

But I had played more than recklessly with fate and with my parents' feelings. I received another alarming cable — 'Your father seriously ill. Do not delay. Mother.' I had friends in my room when I opened it. They knew I had been procrastinating. Many of them had been alarmed that I had continued to dally after the death of my brother. Now they felt I simply had to go and they virtually packed me off. My ticket was purchased and I was placed on a steamer bound for Bombay.

As soon as I boarded the steamer I realized I had been

reduced nearly to a nervous and mental wreck by the strain to which I had subjected myself in clinging to a policy of endless drifting.

The long sea journey reinvigorated me and my mood altered. I spent my time thinking deeply about my life during the past few years. I was appalled by the dream-like existence I had been living, apathetic to all sense of responsibility. England became a faraway haze in a remote corner of my mind, and I arrived at Amritsar with a sense of responsibility to my parents strongly dominant. I was determined to make amends, especially to my father who was ill and whose tenderness to his children, and to me in particular, moved me strongly.

I rushed to the house from the station, ignoring the kindly greetings and exclamations of recognition as I drove through Amritsar in our open two-horse phaeton. I was so intent on heaping care and attention on my father that I was unaware of my lack of attention to the citizens of the town.

I entered my father's room. My mother and a doctor were with him. He lay very still under the bedclothes. In silence my mother rose and held me, her body heaving gently against mine. She took my hand and brought me to my father's bedside. Then she whispered in a voice breaking with sobs:

'Call to him softly, Son.'

'Respected Father. I am here. Gyan ... Gyan is here, respected Father!' I could hardly recognize my own choking voice and I had to stop to control my rising anguish.

His eyelids fluttered and then opened. He looked at me with fixed unblinking claylike eyes, but the rest of his face softened and, very slowly, he spoke.

'Gyan, Son. I was waiting for you. You have come so

that ... so that ... ' He was too exhausted to go on. The words turned into a soft long sigh.

We stayed with him for half an hour before the doctor motioned us out of the room. They told me that the words he had spoken had been his only speech in the past thirty-six hours. His heart was barely functioning, and he could no longer eat. The best herbs and medicines were being used.

'And he will recover now, will he not?' I asked the doctor in a tone of challenge and conviction.

There was a long silence before he replied. 'Anything can happen. We must do our best for him.' He took up his bag and left.

That the doctor clearly felt there was no hope of recovery only filled me with contempt and pity for him. My own determination rose to towering heights. If the others were faint-hearted, what of it, I told myself. I would carry my father to recovery. I would compel things to come right. A demented vigour possessed me. 'Something *must* be done! Something shall be done!' I kept telling myself. I was irked by the obsequious helplessness of the faces of the managers and other functionaries who hung about the house. What do these people think, I asked myself. I will show them what the future has in store — no death and mourning. I ordered them to gather in one of the ground floor rooms. Tell me what is going on, I demanded, and show me your accounts, each one of you. The Master will be up and about in a day or two and what will you have to show him? The men were confused and stunned by my words. But I insisted, and demanded that the next day they give me a full account. As they left I could read their cogitations: was I trying to get everything into my hands before my father died, or was I trying to dispossess any others who might have claims?

Several times that day I peered into my father's room and told myself that he was resting and regaining his strength. That night the doctor remained at our house. Everything possible was done to revive him, but his pulse faded away. I held his wrist and felt it ebb — very peacefully it drifted till it stopped. He was dead.

I gave my attention to the funeral ceremonies. But why are we dealing with the problem of an alien body, I kept asking myself. This is not my father! So stupefied was I, and perhaps such is the nature of faith that I continued to believe that all would be well. I went on fighting. I kept acting and feeling as if the course of events could alter if I tried hard enough. I said to myself, 'They have pronounced a sentence of death. Yes, but all sentences can be changed, just as everything else can be changed. Change is the law of life.' I must change this! That is why I, Gyan Chand, am here.'

My father's body was placed on the funeral pyre, and soon only the shrivelled grey ashes remained. I felt them fill my lungs, my head; and my faith was pushed out of me. But even then, as moments which seem to re-create exactly the very substance, colour, and dimensions of the past, it would come back and make me act as if the present was just a temporary hiatus in the true order of things. At such moments I would be urged forward to accomplishments which would have pleased my father. I even began assuming his manner in my relations with the managers, clerks, and messengers. After blowing up at one of them, I would tell myself that it was ridiculous that I should become a hard and domineering autocrat, but because these poor men cowered before me I decided I was but doing my duty and that the consequences could not but be good. My men too, I felt, knew that I was acting as I did out of a sense of duty, and they obeyed me because it was their duty to do so. In this foolish way I tried to

31

convince myself that it was an orderly and even beautiful arrangement.

But these waves of faith in myself would ebb, leaving me depressed and defeated. The burning conviction with which I had returned home to undo my neglect was, in fact, a piteous mockery. I was fighting for an irretrievable past which, with each moment, was receding farther away, to mock me with its unalterable solidity. Nevertheless, before this feeling of defeat could lead me to turn to some other direction in life, my burnt out faith would flicker again into a struggling flame and I would resume the fight.

Locked in this struggle within myself, I neglected my poor mother. For her the large town house at Amritsar was now like a mausoleum, and perhaps she realized, too, that alone though I was I would need the whole house for the life which now devolved on me as head of the family. She said she wanted to return to the family village near Ajnala and live out her days quietly. Her daughter-in-law and grandchildren were there, and she would expect me to visit her from time to time.

I was in no state to bring her much solace. I said to her, 'Respected Mother, do as you feel you should. I will certainly visit you, and this house will remain yours to return to whenever you should wish.' She left for Ajnala, and my sister left to join her husband in Jullundhur.

I was alone now in the large old house at Amritsar. What was I to do? Money was not the problem, but rather time and the hollowness within me. I could not think of making the management of my properties my major occupation. I was not interested in developing them, and therefore could leave the routine function of maintaining them to the crafty but essentially servile managers who would function with a measure of rectitude because they feared my presence and knew that from time to time I would make surprise visits and inspections. Nor did it materially affect my situation that outside the main gate was affixed my nameplate, reading, 'Rai Gyan Chand, Bar.-at-law'; I had no intention to take actively to the legal profession.

While these stark thoughts were impressing themselves on my mind, in rather an aimless fashion I was exploring the old house. Even in so northerly a town as Amritsar, by the early eighteenth century the rich could build houses instead of fortresses. None the less, on two sides the walls of the house rose a clear unbroken twenty feet of closely packed narrow bricks before they were pierced by the apertures of the ground floor clerestory venetians. Outside the second floor were beautiful old timber balconies, running along most of the east and south faces. On the north, which was the front of the house, was a deep second floor veranda leading out to the terrace over the front porch of the house. The third floor was surmounted by three little hexagonal rooms in the north-west and east corners. At the back of the house was another courtyard, with guest rooms and quarters for the maids and some of the other senior members of the domestic staff. Most of the servants, however, lived in the rooms surrounding the large front courtyard. There were lawns and gardens on the east and west of the house, while at the back was a large patch of waste land.

On the ground floor were several spacious rooms which were little used. They were lavishly furnished without order or taste. There were beautiful old carpets and hangings, some pieces of graceful furniture — divans, chests, tables — and an odd assortment of chairs and sofas. There was a pleasant dining-room and two or three office rooms where my father had met his managers, tenants, and lawyers.

The second floor had been restricted territory. It was my parents' private floor. Of course we often spent time with them in my mother's sitting-room and occasionally in their bedrooms, but we never intruded into the other rooms. These were supposed to be my father's private offices. I found that, apart from the two bedrooms, dressing-rooms, and bathrooms, the floor contained two sitting-rooms, a room full of books, a room for my mother's private temple, my father's private office, and finally a beautifully decorated room hung in old silks and carpeted with an Amritsar rug in rich shades of red, rust, and green. It was furnished with low tables, a wide sofa, and a very large low divan, soft and supple and liberally strewn with colourful silk-covered bolsters. On the walls was a series of beautiful eighteenth-century Ragmala paintings depicting the classical musical themes. Interspersed with them were miniatures of courtly dalliance, fair ladies performing their toilet or in their baths, or waiting for a lover, or listening to music. Though the room full of books had surprised me, as I had not connected much love of learning with my parents, it was this beautiful room that made me ponder. It was another world — far removed from the austerity of my father and the piety and sweetness of my mother. Had I known only the façade of my parents? Was this richly sensuous and beautiful room the real setting for their lives?

To the west the room opened on to a small private balcony. Here, too, was a low divan from which one could just look over the low carved wooden railing of the balcony to the garden and beyond to the western part of the town; and still farther to the green plains of the Punjab. In the late afternoon, as the sun set and the short vivid twilight of the Punjab set in, the scene from the balcony was strangely moving — emptying me into its vastness and filling me with its limpid quickly changing lights.

The next floor of the house consisted of a large number of smallish rooms: the children's bedrooms and their study, and three or four other unused rooms. Above this floor was a terrace — where we had played as children on summer evenings — with the three corner rooms which I have already mentioned.

After a few days, my excitement over my discoveries in the house settled into a quiet stimulation of my natural desire to read. I spent many hours each day in my father's library, poring over old philosophical texts and books on local history. But in spite of this interest I was soon acutely restless. I craved for human companionship and I envied the meanest domestic of the household as he met his wife in the front courtyard and together with her entered their single-room abode.

It was at this time that I first noticed Karamjit Singh. He would come to the house almost each day and spend a half-hour or so in the garden. I gave him two or three days to come up to me. Then I could wait no longer. He responded with extreme reluctance to my overtures. In spite of his reticence I learned a few facts. His father and mine had been very close friends, and when Karamjit's father died — some six months ago — my father had taken an interest in the young man, who became a daily visitor to our house. Though my father had been dead some weeks, and no one had paid any attention to Karamjit, he had continued his visits.

He was a pleasing young man about three years my junior, strongly built, with very gentle eyes and a soft voice which came from rather shy and sensitive lips. He had been very much attached to my father, which surprised me. Most people — and more so the younger generation — had found him taciturn and even awe-inspiring.

In my attempt to create in him an interest in me, I

35

lured Karam into the library and let him borrow books freely. Soon we began to spend several hours each day chatting and eating together. My mood was an expansive one, and Karam was increasingly responsive. But something was lacking. We were always talking about life, and our words would float enticingly before us for a while only to become listless and neutral. I liked Karamjit's gentleness and yet I wearied of it. He was too much like an entranced large animal living in a benign fairy tale.

I tried to work off my restlessness. In the mornings I practised swordsmanship, and was exhilarated by the swaying, lunging, feinting, and quick footwork; then I bathed and read, breakfasted, attended to a few business affairs, strolled in the garden till my midday meal, after which I read again. There were isolated days when this satisfied me; when the yearning for something more slipped away, leaving a joyful response to the beauty of the garden or the sunset glow of the sky, or a feeling of contented participation in the normal events of the day. Even the sense of breaking free from my restless desires was a strength-giving and invigorating pang. At times like this, enthusiastically I would plunge into some new work of philosophy, an old book of poems, or an intricate and delicate tale, and imagine that I was about to enter a new period of harmonious living. But of course I was not really setting out on anything new, and my inner disturbance remained unallayed.

Searchingly, I went to the secret room of my parents on the second floor. Its rich sensuousness delighted me, but something glided away behind the silken hangings in protest against my solitary gaze. Yes, I was desperately alone. But who could I bring here?

One evening I broke out, 'Karam, I am fed up. We meet here day after day. It's been fun, but for god's sake, let's do something else. Anything! What do your people do at this time of the year?' — it was February and far too cold to think of a trip to the mountains.

'Oh, they do idiotic things like going on duck or deer shoots,' he replied, without any enthusiasm.

'Shooting? All right, let's go out for a shoot with them!' My reply took him aback. He looked at me and saw I meant it.

'That's easy enough. My uncle, Daljit, is arranging a duck shoot next weekend on his estate near Patti. They would be happy to have two more guns.' Though he, too, must have wanted something more to happen in his life, he said all this with as much enthusiasm as if we had been talking about an appointment with the dentist.

I tried to explain myself. 'Let's go, Karamjit. It is not the shoot I want.'. I did not really want to go any further. He helped me out by nodding in half comprehension.

It was all arranged. Karamjit and I drove out about twenty-five miles in an open buggy to Daljit's village home. We left early in the morning and got into Jhandasinghwala at about three in the afternoon.

I had expected an old-fashioned village house, but to my surprise Daljit Singh had built himself a spacious bungalow furnished in Western style. He himself was equally surprising. He was dressed, I suppose, in the manner of an English country gentleman of the time — a Donegal tweed Norfolk suit. Entirely unlike his tall nephew, Karamjit, he was a middle-sized, square-built man of about fifty, with a jovial extrovert temperament. He was standing on the front steps of his bungalow when we reined in. He waved us a welcome, his round genial face smiling broadly.

'Come along! Come along, Karam and Gyan. The youngsters of the party are the first to arrive. Good! You can help me receive my other guests.' He gave each of us a hasty pat on the back, with a hug for Karam.

37

'Karam, you young lion! How do you manage it, vegetating in Amritsar? You must come out here more often and develop your prowess.'

Karam smiled. 'My prowess at what, Uncle?'

Daljit's whole face twinkled. 'Prowess at what? What a question! There's no limit to the things a young man can do. You'll see for yourself. And no "uncling" me, either of you. I am D. J. to everyone who comes out to my parties!'

We went in and washed up. Karam disappeared to help D. J. with the guests, but I hung back in our lair. Did I want to be swallowed up by D. J.'s hunting bonhomie? Deliberately, if rather uneasily, I read for an hour and a quarter. I must have dozed off, for it was already early evening when, attracted by the sound of talk and laughter from unfamiliar women's voices, I went to the drawing-room. There were about fifteen persons in the room, including three or four British couples. Their womenfolk were the only women in the room, for in those days our women seldom joined in expeditions of that kind.

D. J. at once hailed me. 'Come along, Gyan. Lady Trip, let me introduce my young friend, Rai Gyan Chand. He has just returned from a stay of four or five years in your country.'

Lady Trip was the overwhelming type. She pitched into me. 'How perfectly wonderful that must have been for you. Five years! Oxford, I suppose; or was it Cambridge? I can tell it was one of the two, mind you!'

I smiled and was about to admit that I had merely studied law at London, but she burgeoned again, 'I *so* long to be back in England! Of course, mind you, I love it here, too. Wonderful people like D. J.' — her eyes tried to glow responsively — 'and this healthy country life. I suppose you spend your time with D. J. — shooting on your various estates?'

Again I smiled, but this time I had to get in a word to interrupt the florid picture of a young hunting and shooting type just down from Oxford or possibly from Cambridge.

'I am afraid this is the first shoot I have ever been to — ' Lady Trip looked incredulously at me. I nodded to confirm my words and went on, 'And I live quietly at Amritsar doing practically nothing.' Her incredulity turned almost to a look of horror, as if she were mixing with some sordid outcast. She turned to look around for help. 'Where is D. J.?' she asked, and added, 'I am sorry. I just remembered that I had to give him a message from my husband, Sir John.'

I spotted D. J. With pleasure written largely across his face he was talking to a young English girl. Around him were four or five of his other guests, smiling and more than a little envious of their host. I pointed him out to Lady Trip.

'Oh yes,' she said briefly, as her keen eye took in the situation. Then she smiled triumphantly. 'Come,' she said enthusiastically. 'You must meet my daughter Jennifer, while I talk to D. J.'

We went up to D. J., who was enjoying himself so thoroughly that he did not notice us till we were practically breathing on his neck and the bright trumpet of Lady Trip's voice had sounded in his ear.

'Jennifer, my dear. This is — this is my young friend who has just returned home after five years at Oxford. You must talk to him while I give D. J. a message from your father.' Lady Trip sailed away with D. J., who muttered to me in Punjabi as he passed by, 'The old sport-spoiling bitch!'

I looked at Jennifer. No wonder D. J. had sworn at Lady Trip. A brunette, blue-eyed, and very pretty in the English manner stood before me. She smiled a greeting,

39

her lively blue eyes lighting up. I cautioned my own surging interest. Perhaps she was just relieved to be rid of the over-interested D. J., or perhaps she wished to talk nostalgically of England. I was mistaken. After a few meaningless words about the countryside and D. J.'s comfortable bungalow, she said, 'I feel terribly out of place here. My father — who will arrive tomorrow morning — and my dear mother love these shooting parties. But they don't interest me — except that I do often enjoy the riding and the country air.' She looked at me, wondering how I would receive this appalling confession.

'This is my first shoot,' I said encouragingly.

'And my last, I hope!' she said. 'We can talk while the others rush around for duck. I feel I haven't talked to anyone for a year! My mother is always whisking me off to things. If only I were allowed to do only what I felt like doing!'

Ingenuously I drew her on. 'Yes, but do your feelings always tell you truthfully? Mine don't!'

She hesitated a moment; perhaps I was only teasing. 'Truthfully yes. Yours do, too. But there's so much else in the way — so much to consider.'

'I agree. It is not a matter of the truthfulness of one's feelings, but whether one can follow them,' I said, this time with more seriousness.

'How is one to know? One isn't allowed to.' Then she smiled again and added, 'At this moment I would merely like to go out and walk in the evening air! Do let us.'

I looked at D. J. Would he let us?

We were walking towards the door. I glanced at Karamjit. He was looking at us, his face flushed and his eyes blazing as if fired from deep within himself. I could not make out what had happened.

As we neared the door, D. J.'s voice pounced from

40

across the room. 'Gyan, do come and explain this point to Lady Trip.' I turned around. He gave me a wink and his lips curled back in a primitive smile of triumph. I looked at Jennifer. She nodded helplessly. I excused myself and went towards D. J. and Lady Trip. He patted me to her side.

'Gyan, you know, is an intellectual. He understands these things better than I do. I will go and see about some of the arrangements, with your permission, Lady Trip.' He bowed, his face beaming, and he walked towards the door.

Lady Trip started to ask me something, and I manœuvred round her so that I could see D. J. He was going up to Jennifer, his arms gesticulating happily.

'Just the person who could help! Come, Jennifer. Let us see about dinner.' And taking her by the arm he walked out of the room, slamming the door behind him. I noticed Karamjit visibly start as the door slammed. His blazing eyes were fixed on the door, and his neck had thickened with anger. The young animal was being roused.

My conversation with Lady Trip was not memorable. It could have been about alligators or angels. She was the sort of enthusiast who could espouse the cause of either with equal zest and absence of understanding. We talked for a while, and then drifted about in the room for almost half an hour before the door opened again and let in first the sound of D. J.'s gay full-throated laughter. Jennifer entered, looking self-possessed and untouched, and D. J. followed, his face beaming for all to see. Two young men in the corner behind us whispered smirkingly in Punjabi, 'The old bastard's made a conquest — like hell!' Karamjit's face was struggling between joyful relief at seeing Jennifer again and the hot anger provoked by D. J.'s boastful laughter.

41

Dinner was announced and turned out to be a noisy affair with amusing undercurrents. Lady Trip sat at D. J.'s right and Jennifer on his left. He ignored Lady Trip as much as he dared, passing delicacies and making chivalrous remarks to Jennifer. Karamjit sat at one end of the table, obviously trying to restrain himself but unable to keep his eyes off Jennifer. His face, superficially calmer now, was that of a man turning over a dozen schemes in his secret mind. Jennifer was responding civilly but coolly to her host. Once or twice she seemed to become aware of Karamjit's gaze, but immediately she dropped her eyes and turned away as if to discourage him. I was the one who interested her. Whenever she felt it safe to do so, she looked at me and smiled. Once or twice I thought she let herself gaze at me. I was flattered and amused. I had neither D. J.'s desire to be the victorious strutting cock by her side, nor Karamjit's consuming feeling — that is what I supposed it to be.

We turned in early, for we were to set out just before five next morning for the lake where the ducks were roosting. Karam and I were sharing a room.

'Well, how are we doing on the first night of our shoot?' I asked. Karamjit growled surlily in reply. I coaxed him. He turned on me, his eyes blazing fiercely. Then he held himself, and shrugged his shoulders. 'I was a fool to come. I knew I shouldn't have. Anyway, let us sleep now as we will be called at four in the morning.'

I fell asleep fairly quickly, while Karam still tossed uneasily in his bed. I awoke as soon as we were called and had to shake my companion hard before he opened his eyes. He had probably not fallen asleep till the early hours of the morning. I almost wished Jennifer had not been at the party, for it was going badly for my young friend.

Karam and I took up a position on rising ground to the

north of the lake. It was not a particularly good spot for duck, but that was a matter of indifference to me, and at least the rising sun would not be in my eyes. All I proposed to do was to let off a few shots every now and again to give the impression of sporting co-operation from our end. Karamjit said he would not fire a shot.

We heard voices behind us — D. J. and Lady Trip. We turned to call a subdued salutation. Jennifer was with them, and they were making for a sort of isthmus in the lake which was the best spot for the shooting.

The ducks began to rise, and D. J.'s gun led off — a kill first shot. Then there were two volleys from his stand — nothing fell. The women were trying their hand. Their failure roused Karamjit. His gun went to his shoulder and he let off both barrels. Four birds went spiralling down. The guns sounded all over the lake, and a few minutes later another flight rose. I dropped one, and Karam again had four down. So it went on for three hours.

After the shoot, we collected for breakfast. Jennifer was dressed rather dashingly for those distant days. She wore a light blue woollen sweater, a canary yellow scarf, and a loose tweed skirt. Her eyes were restrained, but her face was flushed with excitement and yet was pensive in a way. D. J. was telling her who had been scoring the hits. I could see D. J. point out Karam, but she did not seem to want to meet him.

But I had decided that it was only fair to Karam that he should at least talk to her. Perhaps her voice would jar on his ears — not that it was unpleasant, but with an infatuation any shortcoming might spoil everything. I took him up to her and introduced them. Her cool nod struck him dumb. She turned to me.

'The sound of the guns had such an extraordinary effect on me. At first they were fierce and excited, and then they

43

seemed to become just noisy and even rather mournful —
as though the noise did not want to explode the calm
morning air. Am I talking wildly?' She flushed and
smiled uncertainly.

'No, no. Not at all. I must confess I was excited by
the guns when Karamjit' — I nodded at him — 'got going.
He seemed to blaze right into every poor bird. It was
amazing, and he began by saying he would not shoot at
all. Something inspired him,' I added smiling, and
teasingly I asked her, 'I wonder what it could have
been?'

She merely shook her head, giving me a quick look
which said, 'Please don't raise that subject now!' I looked
quickly at Karam. He had flushed strongly, but to my
surprise he also found his tongue. In his slow gentle drawl,
he said, 'It must have been the cool morning air. I love
it, and it clears my vision.'

Jennifer nodded approval and gave him a quick smile.
But again she turned to me. 'Oh! The sunrise and the
gleaming lake with the reflected bulrushes aiming a
thousand spikes at the heart of its loveliness! Somehow ...'
she stopped a moment, and then added very slowly, 'I
felt rather like the lake.'

Jennifer looked up at me, her eyes wide open and
questioning. Did I see the darts that pierced her? I was
perplexed. I wanted to shake my head. But perhaps she
feels she is being hounded by D. J., Karamjit, and some of
the other young landowners who certainly look at her as
if they are appraising a fine young animal, I said to
myself.

'No. You mustn't feel that way,' I protested earnestly.

Now it was she who was perplexed. Her look inquired
more deeply. 'But how can I not? Something in me is
clear and ready — yes, like the lake for the bulrushes.'

I was on wings, but in trouble. She was saying this to

44

me, wasn't she? Had I not left my placid routine at Amritsar because I wanted, needed to fill my life? But this I had not expected. Confusedly, I told myself I must find my feet again — then, then I would know what to do. Fortunately Lady Trip was approaching, her purposeful expression directed at her daughter. Jennifer saw her and gave me a glance of helpless release. I said I had to thank D. J. for letting me come to the shoot, and with that I slipped away.

D. J. refused to receive my thanks. 'Thanks? Whatever for? If you are enjoying yourself, that's fine. If you are not, tell me what I can do about it. And in any case this is just the beginning. There's partridge shooting this afternoon and hare coursing tomorrow. And then if everyone were not so hard-working I would suggest a trip to another lake, twenty miles to the west, to shoot up the duck there.'

He talked on affably, but I heard very little. I was glancing at Jennifer. Karam had joined her and Lady Trip. He was talking to Jennifer but I could not hear what he said. When she replied she did not look at him; and once she shook her head. That pleased me. She looked lost, I decided, now that I had left.her.

When we got back for lunch, Karam and I were together in our room, changing from the heavy clothes we had worn for the morning's outing. He looked gloomy, and fumbled about the room.

'Cheer up, Karam. You did some wonderful shooting.' I said breezily.

'Shooting? Who can't shoot with a gun! But how does one say anything that goes home with that Jennifer girl. Why does she hate me so? What have I done to her?' His face was flushed with dismay.

Poor man, I thought. Of course, he had done nothing except look at her darkly all the time he was with her so that she felt any moment now he might pounce on her. But how to tell him this?

'Listen, Karam. Just let yourself relax and she will talk to you as she does with anyone else.'

'Anyone else!' he practically yelled at me. 'Does she think I am just anyone? Can't she see I think she is a queen of beauty? Can't she let me adore her from not too far off? Does she think I will bite her?'

'Yes,' I replied — it came out automatically, and it was just right. Karam burst out in a pleasant roar of laughter.

'Oh, silly fool that she is! And I suppose I, too, for frightening her. I suppose I do look pretty frightening when I am not getting things my own way. You are right, Gyan. I must ease off. Stupid of me.'

We went to lunch. Jennifer was already surrounded by young men, while D. J. was being held off by the intrepid Lady Trip. Karam dashed up to the table, and filling a plate with some tasty appetizers he took it to Jennifer, looking as bland as he could. She thanked him winsomely. Inordinately pleased, he thrust himself into the group surrounding her. His crack shooting at the lake marked him out as the hero of the occasion. Blatantly he used this to strive for the precedence with her that he so desperately wanted.

'Miss Jennifer, you must let me give you a shooting lesson this afternoon when we go out for partridge. Of course, they are more difficult than duck, but that will only make it more fun. They rise like bullets and not in large flights. It's pretty awful to cut off their speedy exit by slapping them with shot, but the precision of it gives one a wonderful feeling.'

The others fell back a step or two. This was clearly Karam's field. Jennifer was enthusiastic. 'Do teach me. I will look out for you. Perhaps my father will expect me to be at his side — but maybe not. I would only be in his

46

way. He loves partridge shooting. Though, between us, you could probably give him a useful lesson or two. He is not terribly good at it.'

'No, no. I couldn't possibly teach him,' came back Karam's quick response. He did not want to be landed with Sir John.

They both laughed, and all seemed to be going well. At lunch I avoided her, even though she hailed me twice as we were going around the buffet table; but I kept my eyes on her. I decided that my withdrawal was upsetting her a bit, and I was secretly pleased that this should be so.

In the afternoon Karam was delighted as we set out after partridge. When we lined up, Jennifer was at his side. I was on my own about fifty yards to one side of them. The beaters got going. There was a long call of warning from a male bird and then three singletons rose with the desperate singing rattle of their wings. They were out of range for Karam and Jennifer but one was near enough for me to take a chance. I fired and down it came. I saw Karam look at me in surprise. Jennifer waved to me.

We moved along the field. There was a rise near Jennifer and Karam. He got her to try, but she missed. The bird was already far out, but Karam followed and got her. It was an amazing feat of marksmanship. The shoot went on. Jennifer — according to Karam — nearly hit three birds. His shooting was superb, and to everyone's surprise, including my own, I had only one bird less than he.

When we returned in the late afternoon, to refresh ourselves, Jennifer came up to me with a cup of tea which she pressed me to take. Then she brought herself one and said to me, confident that she had me in the mood to talk to her again:

'Why didn't you offer to teach me? I am sure that with

47

you I would have brought down a poor bird or two. I am frightened of your friend. Besides, one can only learn with certain people — people whom one — ' she stopped.

'But why be frightened of Karam? He's a wonderful chap and wouldn't hurt anyone. And I couldn't teach you. I can hardly shoot!' The last point was not too obvious and I suppose I said it to draw her out.

'I know you can't!' she said teasingly. 'So each time you let the others have the first chance — and still you shot as many birds as Karam!' She stopped. Then she went on, the pitch of her voice dropping. 'But don't you see, only you could teach me — anything.' She looked at me, her eyes softening.

I turned away, and said out into space, 'Anything sometimes means everything, sometimes nothing.'

She was looking straight ahead when she replied, 'In this case it doesn't mean nothing, so I suppose it means the other.' She spilled a little tea at the end of the sentence.

I turned to help her. Her face was very calm and thoughtful. We mopped up the tea.

'There,' she went on. 'All is well again. But life is too difficult. Tomorrow I go to Lahore with my parents, and you will return to Amritsar. Lovely word, "Amritsar". What does it mean?'

'The lake of immortality,' I replied.

'Amritsar, immortality. Elsewhere nothingness. That may well be how it is.' She smiled, and before I could respond by word or look she had got up and gone to join her father who had just entered the room and was survey-ing it with the sort of dour inspectorial look that any important chief of police — which he was in the Punjab — is duty bound to develop.

She left me thrilled by her directness and by the tenderness of what she had said and implied. A tingling

48

sweetness flooded and intoxicated me. But what had this girl meant? I glanced at her again as she talked to her father and a group of D. J.'s friends. Jennifer was laughing happily, and her blue eyes seemed to reflect all the light in the room. I smiled to myself. Perhaps she was the sort of person to whom words came easily, and I would be a fool to let myself be over-affected by them. But still the sweetness surged up in me. Annoyed by its excessiveness and persistence, I got up and went out seeking the calm of nature.

I roamed through the garden and into the fields. The spicy scent of the earth and the deep green fodder crops conspired with my inner intoxication. It was as if I was hemmed in. I sat down to think my way through all this. Absurd! I told myself. Wonderful, came the inner answer. You are Gyan Chand, on holiday from Amritsar for a weekend, but rooted in your life there. No, you are creating from the past — the past of your dead father, your mother who has retreated to the village, and the tradition that dogged your steps while you were in England. You are free at last.

The earth and the fields stretched around me — firm and unchangeable, and pouring into me their own intoxication — sprouting, scented life, not caring for anything except its own sweetness and strength. But my mind went on churning out its leaden scepticism and caution.

I got up and steadied myself. Tomorrow, Jennifer would leave. All this confusing intensity would pass and I could work out my life more reasonably. Unelated by this conclusion, I re-entered D. J.'s garden. It was just after sunset and the light was deceptive. But a feminine figure was rapidly walking towards me. It was Jennifer.

'Why did you go out without me?' she asked simply — as if I had broken my word to her.

God knows I had not been without her for a moment. 'But I wasn't without you,' I blurted out.

She nodded rapidly. 'I know. That is why I should have gone with you. In the room you left me with your shadow.'

You are free, free, hammered through my temples. I took Jennifer's hand and she pressed mine in response. Then she smiled, saying, 'Would they miss us if we fled to the Lake of Immortality? Let us go away at once!' and she tugged at my hand playfully.

'But, alas, the lake isn't mine!'

'Let us capture it together, Gyan. ... But why am I like this with you, saying absurd things which at the same time seem so necessary to me?'

'No, no!' I cried quickly. 'There's nothing absurd. You are only saying what is being said all the time inside me. You are talking for both of us.' I stopped, surprised at what I had said.

'I know that already, Gyan. It is as if the parts are already written out for us. And I know there's more to come.'

She pressed my hand, and made a quick movement so that for a moment all her weight seemed to lie in the palm of my hand.

'I must go and change for dinner. Any minute now and they will be looking for me. Please, please don't forget your lines in the play.' And saying this, she walked quickly towards the house.

Though D. J. had intended otherwise I learned later that Jennifer had asked to sit by me at dinner, saying she wanted to talk about Amritsar. She was in a gay humour and contributed liberally to the lighthearted chatter initiated from the centre of the table by Lady Trip and D. J. Even the dour-looking Sir John relaxed, and the controlled drawn lines on his face became ordinary human wrinkles. Karam alone was out of it all. His black looks deepened as the dinner progressed. I tried to draw him into the conversation, but he was having none of the general gaiety.

It was the Trips' last evening. Jennifer asked me whether I ever went to Lahore. 'Very rarely,' I replied, explaining, 'It's much too official a city for me' — which was true. My family had kept itself aloof from official entanglements. Consequently, at first we were in great disgrace with the British Government and were even suspect. My grandfather had been a prominent feudatory under the Sikhs, and the British — when they took over the Punjab in 1849, confined him for five years in an old fortress in Central India. On his return to Amritsar after his release he never referred to that five-year period. It had wounded him deeply that the new power in the country, instead of respecting him as an honourable foe, had swept him off to Central India and cut off his liberties for five years, and for ever after his indignation silenced him on the subject.

I wanted to tell Jennifer some of this but there wasn't time. I said only a little, and that rather obliquely. 'We feel much more ourselves, completely apart from the official world. For people like us who would never be able to submit, it is best to keep out of the way of the Government.' I stopped, not wanting to upset her and wondering how she would take what I had already said.

'But I entirely agree. You know, I think it absurd that we English should be in your country in the way we are. I hate it. The only time I enjoy anything is when we are like this with your people. Then I forget that my father is an important official and I feel we are friends from England who are visiting friends in India.'

I looked at her face. She was being absolutely sincere. But she was young, and the shadow of how she could feel several years later when she herself might be the wife of a colonial official fell across my eyes, making me wince.

Jennifer behaved as though the rest of the company in the room were in a world outside hers. Apart from brief sallies she talked only to me, telling me about her childhood, her longing to escape from the round of official events and sociability, and her feeling that here in India there was something she could pursue. She pressed my hand very warmly when she said goodbye. I would not see her again as the Trips were to leave by car at about six o'clock the next morning.

When we turned in for the night, Karam could control himself no longer. 'Gyan, I never expected that you would join with Jennifer in slighting me tonight,' he said, with the veins in his neck and forehead throbbing.

'Whatever can you mean, my dear Karam. I kept trying to draw you into the conversation. What more could I do?' I asked quietly and casually while I continued undressing.

'How can you say that! You kept the girl by your side the whole evening and gave her no chance to talk to anyone else. She must think you a boor,' came back his absurd reply.

But it was a relief to me that he thought it was I alone who had been taking the initiative. I do not know why it made me feel better; I suppose I enjoyed keeping to myself the new delight I was experiencing. I decided to leave his conception of the matter undisturbed.

'I am sorry I kept her from all the others. Well, as you know, she is leaving tomorrow morning, so that there is not much we can do about it now.'

Karam went to bed without bidding me good night. I called him but he did not reply. I put the matter off — I would talk to him the next day when his rage would have subsided. But my procrastination had a deeper cause. I wanted to revel in Jennifer and the problem she had created for me. I was fully aware of these two irreconcilable aspects of the situation: her beauty which bound me to her, and the maddening fact that I could not possibly think of a life with her. But, as I lay on my bed, even my mind refused to fight against the urge to go to Jennifer, to hold her in my arms and, defying tradition, custom, the compulsions of geography, the political situation, and the differences of religion and race, to make her my wife. That night, it all seemed so simple — so inevitable — and I fell asleep smothered by the sweetness of this fantasy.

I was calmer when I returned to Amritsar the next day, and the ecstasy over Jennifer seemed unreal in my old surroundings. But I was determined to put it to the test by seeing her there. Immediately I wrote to her, begging her to pay me a visit. There was no answer, and as the days passed I began to feel I had behaved foolishly.

For several days after our return to Amritsar, Karamjit did not visit me. Then he came one day and was his normal, rather stolid self. He looked at some books in the library, and later in the evening he announced that he had decided to get married and was about to ask his mother to arrange a match for him.

This unexpected news quickened my longing for Jennifer and for the tenderness of love. 'That's a big decision,' I said to him. 'I wish you well, Karam. I must confess it's a matter I have been evading for myself.'

In a way, I was impressed with the news. His people were liberal-minded enough not to press him to marry just yet. He had decided the matter for himself.

Slowly, almost painfully, he explained. 'You see, when I saw that Jennifer girl, it did strange things to me. I felt the pace of my blood quicken; it seethed in me and took control of all my faculties. I know I hit out at you. But I was not really responsible for what I was doing. This

53

passion for a woman is utterly maddening. And look how arbitrarily it attacks one! I did not know this girl. She was pretty, yes; but one sees many beauties every day in the streets of our town. I suppose it was just that the general atmosphere of the place and situation was conducive to an outbreak of lust in me which got right on top of everything.'

'So now?' I asked, glad of his frankness.

'So now I will marry some beautiful girl, and I will be calm and reasonable in my dealings with the rest of the world. I don't want to get that mad again!'

I smiled at him, but I really had no right to be sceptical. Besides, I envied him. Why not take the decision to marry and live my life out with the beautiful and delightful girl my mother or sister would choose for me? What he had said could be very sensible.

I shook my head as if to deny the effect of my smile. 'No. No,' I said quickly. Then I went on, 'I hope it will work just as you say. It certainly could, Karamjit, specially if you feel that way about it.'

He nodded. His face was very calm. He seemed to be sure of himself. The next few days passed without either of us referring to his decision. Then casually he told me that his mother had found a lovely girl for him in Jullundhur. He had asked to meet her but that had not been conceded. In fact the request might even wreck the match, but his mother was trying to assuage the wounded feelings of the girl's parents. They had been shocked that Karam had wanted proof of their assessment of their daughter's looks and charms, which had been accepted by his own mother who had met the girl several times. Karam added that he had decided to accept his mother's judgment and to marry her if it did not now fall through because of his request. He seemed happy at the decision, and as he left that evening he smiled warmly at me.

'Gyan, you are a couple of years older than I. It is high time you married. Why don't you?

I tried to smile, but his words created in me a curious feeling of being left behind: as if I were letting life pass me by.

On the very next day my gloom broke — there was a telegram from Lahore: 'Coming to the Lake of Immortality on the 17th early afternoon. J.' Coming! ... for ever? No, Gyan. Her parents must be coming on a brief official visit and she will accompany them. She might see you for a minute or two. Nothing else is possible — how could a young English girl come alone and visit me at Amritsar? Not that I would mind the risk of gossip. I am alone and unmarried. Besides, as a professional man, a barrister recently returned from England, anyone might come to me for legal advice. But Jennifer — she could not possibly come without her parents.

The next day was the 17th, and at about two in the afternoon a closed phaeton drove into the front courtyard. There were a few of these formerly imposing-looking but now antiquated two-horse carriages which plied for hire in the town, and their stand was at the railway station. I realized then that Jennifer had come alone. The coachman halted his horses and climbed down to open the door of the phaeton. Jennifer stepped out.

She glanced around the courtyard. I could see that she was bewildered by the sight of this large old North-Indian town house. She had probably assumed that, like the more Anglicized Indians, I had a bungalow in the quarter of Amritsar where the British lived. I went down the steps. On seeing me, she was immediately herself again and came towards me smiling. But her face was strained and pale. Had she decided to escape for ever from her small official world?

'I didn't know that I was coming to a medieval castle,

Gyan. This is even more adventurous than I thought it was going to be.' She said this in a mixture of eager interest and determined fearlessness. But I knew from her expression and tone that she was paying me only a brief visit.

We walked into the house. 'It isn't a medieval castle. It is just an eighteenth-century town house,' I said as flatly as possible to conceal my excitement and the pain of deprivation, now that I knew she would not stay.

She was looking at the high ceilings and the baroque flower-patterned frieze ornamentation on the walls. 'You see, it is not severe enough to be a castle. It does not really provide the extreme conditions required for high adventure,' I went on. 'But tell me, how did you arrange this? Do you know it probably hasn't happened before to a young English girl? And what did you tell your parents?'

She half sighed, half laughed. 'I was determined to come and see you. It sounded so reasonable to me. My parents had to go down to Delhi — my father has been called to an important conference, and then he's going on to Calcutta — the Viceroy wants him to be the head of the Intelligence Bureau. Well, I am to join them in Delhi. Amritsar is on the way. So it seemed reasonable and friendly that I should break journey to visit you.' She wanted me to say that I agreed.

'Reasonable? Logical, yes, but most unreasonable for you, my dear!' I said smiling at her, and speaking as casually as possible.

'I know. After I had worked out the clear logic of it, I turned to the practical arrangements and it all seemed impossible. First they insisted on sending down a police guard with me from Lahore to Delhi. I stopped that by calling my mother. She got Father to turn it down. Then reservations were made for me straight through to Delhi on the night train. I decided to let them do that. My parents were wired my time of arrival. I went several hours earlier to the station, and keeping my reservation on the later train from Amritsar to Delhi took an early train up to here. I had to evade the servants and friends who took it as a matter of course that they would see me off.' She looked worried, but shrugged it off, adding, 'I will wire them from Delhi tomorrow telling them I am with my parents, and then all will probably be well!' She sighed, and suddenly turned quite pale.

I brought her a glass of brandy and soon she was relaxed and cheerful again. I put my arm around her to let her feel that she was protected and safe. She looked up and smiled. 'Tell me, Gyan, how are you? You look very well, and terribly handsome.'

I turned my face away. I was not used to my personal appearance being discussed, nor could I bring myself to compliment her on her own beauty.

'Come, Jennifer, let me show you this house in which I live alone.'

'Not alone?' she queried.

'You will see for yourself.'

She smiled a little at the oddness of the furniture on the ground floor. But my mother's sitting-room on the next floor she thought was very beautiful. The whole floor was warm and well-styled, she said. Then we went to the next floor where I had my own rooms. She laughed at me when I led her into them. 'You live like a monk. The walls are bare and there is so little furniture' — only a bed, a cupboard, a chest of drawers, a sofa, and a small table were in the room.

'Yes, but look at the view!' I said enthusiastically. She went to the east and south windows. To the north-east, in the distance were the faint outlines of the Himalayas against the soft blue haze of the horizon. In the foreground

57

were heavy groves of mangoes, pomegranates, and other fruit trees, and in the centre the endless steppes of the Punjab with the young wheat brilliantly green, and the sombre giant wild plum and gaunt babul trees dotting the fields. To the south the distant landscape was harsher but wide open after the section of the town which lay below us. In both views there was a great sense of space which almost entirely took away the feeling of being in a room.

'You see what it does to the room?' I asked.

She came back to the centre of the room and then she nodded. 'Yes. It is wonderful. But Gyan, it does leave out something — the warmth you must need!' And she gestured with her hands, half pleading with me to do something to rectify the omission and half offering her own warmth.

'But, Jennifer, I do live alone. Better then that my room should be like this, don't you think?'

She was restless and disturbed again. We went up to the top terrace and to one of the little hexagonal rooms.

'I would love to spend all day up here. It is lovely — but then I would not want to go down to that severe bedroom of yours ... forgive me, Gyan.'

I smiled. 'There is no reason why you should, Jennifer. Let me show you where you could go,' I said cryptically and glanced at her wondering face. 'Come,' I said, taking her hand.

I had not shown her the secret room on the second floor near my father's rooms. Now we went down and I led her into the library, my father's sitting-room, and his bedroom.

'Yes,' she said nodding. 'This is better. I *could* come here, if you came with me,' but her blue eyes had not lit up. The rooms were acceptable but had not kindled any enthusiasm in her.

I then led her to the secret room. The wide doors to the balcony were open, and in the light at this hour of the afternoon the whole room acquired the intricate loveliness of an Indian scene on an Arras tapestry. The colours of the silk added their lustre to the light now declining from its noontide clarity to a tinge of faint gold. She went up to the Rajput miniatures on the walls. They were like magic mirrors, each changing the reflection of the room into their own mould of exquisiteness. And then she went up to the balcony, admiring its old carved woodwork and the highly polished walls of white and deep orange-lime cement.

She came back from her tour of quick inspection, already physically raised to a tiptoe of elation, and without a moment's hesitation she threw her arms around my neck, and drawing very close she whispered in a voice resonant with an opened inner spring of joyfulness:

'I knew. I knew this, really. I saw it from the first in your gentle sensitive face. Even when your eyes were often distant, it was as though deliberately you had withdrawn, but never into barren coldness. You are as I thought you were, only much more so. I knew I had to come here to you!'

She held me closer, and my arms were around her now. The warmth and vibrance of her body, her words, her coming to Amritsar, every gesture opened something in me. What she had detected was true. Often at D. J.'s when she had looked at me, or even when we had talked, I had held back from her; deliberately, I had remained at some distance from her. But on the last evening at D. J.'s house I had yielded some of the intervening distance to her, and then across most of the remaining space she had drawn me by the frankness and warmth of her feelings and by the surge of her whole personality. Now nothing remained to separate us.

We went to the divan and, in the profusion of our embraces, I felt her melt into my being. There was no struggle towards the full expression of love. Our feelings drew us into it and beyond to a tender quietude.

Later we realized that the light from the open doors was painting the room in deep gold and lamp black. There was immediately in both our minds the rude intrusion of her carefully laid plan to take the night train on to Delhi, and the spell was broken. We sat up. It was she who spoke.

'It may be just sleep-walking, or it may mean nothing at all. How are we to know? But so truly this is good, Gyan. This being with you. The fact that I was born in Sussex, went to school in London, and have lived with my parents in the stupid seclusion of a British colony in the midst of your country — all this is like nothing. I feel you all through my being, and this place of yours — medieval or eighteenth century? — is the reality of the earth for me. What shall I do now? Tell me, Gyan!' And she shook her head, contradicting her cry to me. She did not want me to tell her, for the beauty of the senses, of feeling, of words, of place between us, could be but ephemeral. And nothing I could say could alter that fact.

I held her close with one arm, pressing the fingers of my other hand into a bolster. There was no point in trying to think the thing out. Thought was really starkly clear in that moment of high feeling. It is not that thought is clouded by feeling. Only, often in such moments thought is bitter as gall. But neither Jennifer nor I were the kind who would push it aside. The gall, then, and ourselves.

We got up. She went to my mother's dressing-room and freshened up. Then we went downstairs and I ordered dinner, for the night train did not leave Amritsar till nine o'clock. We sat in a small sitting-room attached to my ground floor office and talked.

'Please do tell me about yourself, Gyan,' she asked, begging me to unravel for her my thoughts and feelings.

I wanted to tell her with all frankness. But I knew only my past and I had a hazy dream for the future, but of the present — which was what she wanted to hear — I really knew nothing. Is not that how it too often is? But I struggled to grasp the present and put it into words.

I told her of the incident in my childhood with the elephant, and of some of the things I had done since then leading up to my present state.

'You see, by accident I won my freedom at the age of seven. And then I had to fight to regain myself, which I thought I was succeeding in doing by returning home from England. But the death of my father and the departure of my mother have deprived me of what seemed to be my new road. Now I do not really know.'

She drew nearer to me on the sofa. 'No one really knows where he is. And because you do have more freedom than most of us, for you it is more difficult to know. And when we assert our freedom we come to the farthest reaches of life — wonderful rare places, but impossible to possess — like our present love. I know, I think, what will happen to me. My parents are now going to be at the headquarters of the Government of India where there are hundreds of young Englishmen struggling against their loneliness. Many will be attracted towards me. My parents will encourage the best of them and I will be expected to agree to marry.' She sighed and stopped for a moment. Then she went on. 'And I suppose it will happen eventually, drawing me into the old grooves — perhaps not entirely successfully.'

We were silent, our hands pressed together. There were sounds on the cobbles of the courtyard. I got up and went to the front veranda. Karam was approaching the house. He had not seen me, and I just had time to slip back to the sitting-room and tell Jennifer that she should not let herself be upset by him. A shadow fell across her face but in a moment she responded to my smile, nodding cheerfully. I went out to receive Karam.

On the front veranda, I said to him, 'Jennifer Trip is on her way to Delhi. She has just come in to see us before taking the evening train.' It was not entirely the truth. I watched his face eagerly to catch his response to this announcement.

Slowly his eyes lit up. Were they going to smoulder in rage or resentment? It was the light sparkle of pleasure that filled them. His smile widened to a broad grin of delight.

In the sitting-room he greeted Jennifer very pleasantly. I was relieved, and soon we were chatting easily. Karam's pleasure seemed to me to be quite disproportionate to the situation. But he could not contain himself for long.

'Miss Jennifer,' he said, sitting up straight in his chair — an unusually vigorous gesture for Karam — 'I learned a lot about myself from meeting you. I am deeply grateful to you. And now I am getting married. I learned that this is what I should do. You, too, should marry — and even Gyan, though about him I am not sure.'

Jennifer laughed, partly in thankfulness that all was well with Karam.

'But why should I marry?' she asked of him. She stopped a moment and then added in a deeper tone of voice, 'And why not Gyan?' She had tried to separate the two thoughts which in her wishes were one.

Karam did not sense the wishful fantasy in her mind — which she had almost spoken, to stir again the depths of pain within me. 'I don't know why I said quite that. But Gyan is able to cope with his turmoils through his

62

own inner strength. And perhaps this is true also of you, Miss Jennifer. I don't know.' I looked at him, wondering how I had given him this impression about myself.

We sat down to dinner. The servants had prepared bland dishes so that Jennifer should not have to battle with the full sharpness of chili and spice. She enjoyed the strange viands and vegetables and said she could eat this food for ever.

We drove her to the station and saw her train pull out. She waved sadly till the night blotted out vision between us. I dropped Karam at his house. He was still in high spirits, but as for me, aware that this beautiful episode had been cut off for ever, I returned to my lonely mansion.

Part III

FOR days I lived in the world of beauty and tenderness that, together, Jennifer and I had created. I was dead to all else except that which, either arbitrarily or by obvious association, seemed part of that magical self-sustaining world. On those spring mornings, whenever the breeze brushed past, I awoke to Jennifer's touch and to her words and the vibrancy of her voice. More directly, the faintest whistle of a railway locomotive was like a warning of parting. Karam, who had been associated with Jennifer and myself, and because of that might have been expected to fit into my closed world, was an unwelcome intruder. I immediately retreated within myself when he visited and had to take myself strongly in hand in order not to let my resentment get the better of me.

For months after Jennifer's visit I wrestled with a painful and even alarming existence. I struggled against being swept away by fierce inner surges which strained to the breaking-point every fibre of my being. When I relaxed, my dismal confusion regarding the future would be transformed by the limpid call of desire to be with Jennifer. Like a light she shone before me. Why not sweep aside tradition and inertia and go to her with my vision clear and with faith in our future together? We could give each other happiness, we could travel, sit still, lie locked in each other's arms; and looking into each other's eyes, we would delight in life. Why, why could this not be? Yes! It could be! It could be! And for a few moments I would be elated by these thoughts. But something would remind me that Jennifer, for all her courage and spirited mettle, had said that our love was impossible to possess. How then could I go to her now? The elation would deflate and I would be sucked back into a maelstrom of depression and confusion. I do not know how I struggled through this period to a viable calm.

I clutched at anything that came my way. Again I welcomed Karam's visits. He was preoccupied with his coming marriage and all that it would do for him. He talked as if he would be called upon to perform feats of heroism in his new duties as a husband and a father. I would smile at his wildly enthusiastic words, and yet, I would find them goading my mind. Was I losing my way and lagging behind in the adventure of life? Without meaning to do so Karam stimulated a sharp and unpleasant restlessness in me.

'It is a pity,' he would say, 'that you, Gyan, are not the kind of person who wants to increase his responsibilities. You are content with the quiet and easy life you lead. I am just not like that!'

Though inwardly smarting, I would smile and explain rather lamely, 'It isn't that I am not the type that would not welcome further responsibilities, Karam, but you see ...'

He would laugh his deep, slow laugh. 'That's just it, Gyan. On balance, you don't want any more responsibilities, whereas I do. Well, perhaps you will change.'

It was humiliating to be patronized thus by a younger man who I also regarded as inept in the art of living. Unwillingly, I let a new train of thought develop. I was forced to admit to myself that it was absurd to go on existing in the exclusive world of my memories of Jennifer, or in my earlier, and by now totally frustrated, momentum of rendition to my parents. I must learn even from Karam.

If he could fall back on marriage and enjoy the prospect, at least I could lay myself open to it. I decided to visit my sister at Jullundhur, and wired that I would arrive there in two days' time.

Though Jullundhur was a bare fifty miles from Amritsar and was the smaller of the two towns, it seemed to have attained a quicker pace. It may be that the fact that adjoining it was a large cantonment of British troops, and that in other ways it had been subjected to more outside influences than had Amritsar, had urged it to a kind of assertion of its own vigour and capacity for urban living.

My sister, Shakuntala, drove herself about the town in her own buggy, and was quite a star at the local tennis club. Tennis was a new game for the people of Jullundhur and they had taken to it 'just to show the British cantonment'. When I arrived there was great excitement because a young girl named Amrita had won the local championship, defeating all the women at the cantonment.

Shakuntala was delighted to have me at Jullundhur. 'Gyan,' she said, 'you will be a hit here, after your three years in England. I have some ideas for you.' And she gave me a conspiratorial look.

She was so much more self-assured than I remembered her in Amritsar. I smiled and shook my head in response to her look. She nodded as much as to say 'you will see'.

She took me about in her buggy, introducing me to her friends, and tried to get me to join in the life of the club. My foreign travels were a cachet and made me a man of special sophistication to some of the younger people at the club. All this showing me off gave Shakuntala great pleasure and perhaps added something to her own standing at the club — I decided that this was all her conspiracy amounted to. But I was mistaken. She had deeper plans for me. They also concerned Amrita, the tennis champion. This young girl, too, was on display. When I arrived at Jullundhur, she was at Delhi being fêted by relations, but before I had been at my sister's for a week she returned to the town. That very day my sister asked her to come to tea before she went on to the club for her game of tennis.

Amrita was a handsome athletic-looking girl. She had a strong face, rather narrow, a fine aquiline nose and keen dark eyes. Her mouth was full but somehow did not suit her narrow face. She expected some notice to be taken of her.

My sister introduced me. 'Amrita, this is my brother Gyan ... I'm sorry — Rai Gyan Chand, barrister-at-law!' she added with a little laugh.

'And this,' she said, bowing to Amrita, 'is our tennis champion, Amrita.'

Amrita stood very straight and raised her hands together in greeting. I did the same as politely as I could, realizing that my sister wished me to acknowledge the presence of Amrita in a special way.

That did not take us very far. We tried talking tennis, but I was an indifferent player, neither knowledgeable nor enthusiastic. I tried her on riding and fencing, but she was strictly a town-bred girl and these forms of sport did not interest her.

We went on to the club. I admired Amrita's tennis — genuinely, for she played with style and vigour. Later we dropped her at her father's house. Shakuntala then told me that her father was a prosperous doctor in the town, who had amassed quite a fortune and had bought large town properties. That did not interest me. Shakuntala asked gently, 'Don't tell me you are prejudiced against her because her father is a doctor!'

'No. Of course not. At least, I do not think I am,' I said honestly enough, trying to be certain that I did not have any of our usual prejudices against people in the

67

professions or in business. You know how it is with our people. They think the only honourable ways of living are as landowners or in an executive government job — no matter how humble it might be. All else is just commercial or trickery.

For the next few days Shakuntala persisted in arranging meetings with Amrita. I did not know whether she had taken the girl into her womanly conspiracy, but I became aware of Amrita's interest in me. She smiled at me from the tennis court, she asked me about Amritsar, about England, and what I did with my time. But though I tried to respond to her, mainly for my sister's sake, I just could not. She did not interest me at all.

I was becoming acutely uncomfortable, but I did not want to leave Jullundhur. And yet the situation was doubly frustrating. First, I did not want to continue the dreary programme of meetings with Amrita; and then, again, I had to admit to myself that the main reason which had brought me to Jullundhur was that I had hoped to raise with Shakuntala the matter of her arranging a marriage for me. Shakuntala had anticipated my purpose, but in the particular person she had chosen she had confronted me with the impossible.

But my sister was a sensitive person, and seeing that I was making no headway with Amrita, she discreetly took to her bed for a day or two and broke the continuity of our visits to the club. Then she talked to me quietly, her eyes thoughtful and solicitous.

'Gyan, I do not want to push you into things. I simply assumed that after your period in England you would like going to the club — the British at the cantonment next door seem to do nothing else! But do just as you please while you are here.'

That set at rest the matter of Amrita and re-established a quiet understanding between Shakuntala and myself, which in a day or two became so mutually responsive that I was able to tell her of my restlessness. She realized that it was a deep malaise by now, and knowing my personality she nodded assent when I said it might be good for me, to marry a quiet, gentle, beautiful girl. I would not want to go through the process of approving of her, much less of falling in love with her.

'Are you sure that is so?' asked Shakuntala.

'Yes, it is so, I believe. I would rather marry as a duty. I trust your judgment now that you know I am not looking for a tennis champion! I am not being unkind to Amrita, but as you know now she is not at all the type of person for me.' Because I thought Shakuntala would want me to be explicit about so vitally important a matter, I continued, 'I want someone who will let me take the responsibility for her, and who will, in her way, do the same for me. You know how a man is utterly lost without the tenderness of a woman.'

I could see that Shakuntala now realized fully what I meant, and that she regarded my assignment as being fitting for a mother rather than for a sister. I stayed on at Jullundhur for a few days, quietly.

When I returned to Amritsar I no longer felt irked by Karam's references to his impending marriage. And just before it took place — about four weeks after my return home — I had a letter from Shakuntala telling me that she had found just the person I wanted. I did not break the news to Karam, but during his wedding festivities he said to me, 'Gyan, you look so happy — almost as if you, too, were getting married!'

After Karam's marriage I hastened to Jullundhur to meet the girl's parents. They had still to meet me and to decide whether I would be acceptable as a son-in-law. My sister drove me to the village of Roopwal, about eleven miles north of the town. My parents-in-law to be had a farm there, though their main properties were in the two neighbouring villages. I fell in love with my future mother-in-law. She had a wide generous face, and kindly large eyes which were very expressive, especially when she talked. I was happy that she should be the mother of the girl I was to marry. Later, on our way back to Jullundhur, when Shakuntala said, 'Basanti is so like her mother, and I saw you liked the old lady,' I was filled with joy at the announcement.

There was no reason to delay the marriage. After consulting the horoscopes and the family priests a date was fixed about five weeks ahead. When the time came, we went in quite, a procession of phaetons, traps, bullock carts and on horseback from Jullundhur to Roopwal. Karamjit and his wife were in our party. Roopwal was much more marriage conscious than a town. The whole village was caught up in the mood of gaiety, and the village musicians played with a zest and skill at improvising which I had not heard at Amritsar. Early the next morning we returned to Jullundhur for more festivities at my sister's home; and after a day there we went on to our house at Amritsar for a final evening party, dinner and music till dawn.

During the marriage festivities I caught three brief glimpses of my wife Basanti. She pleased me in the way I had hoped for. Of course, I was disposed to be pleased but there was no doubt of her loveliness. Basanti, like her mother, was wide of face — the sort of face which invites confidence — with dimpled cheeks, large dark eyes, and a playful mouth. Her brow, too, was wide and emphasized the strength and calm of her expression. It was the face, too, of a mature person and I kept recollecting with surprise that she was barely seventeen. She was of medium height and moved very gracefully. It never occurred to me to wonder about her figure at that stage,

70

which in any case was so overlaid with her wedding finery that it would have been a frustrating speculation.

As the dawn broke our wedding guests were departing — except those who were staying at the house. The fatigue of the occasion pounded in my head and weakened my joints and limbs, and I could see that Basanti was hardly able to keep standing. We spent our first day of married life fast asleep in our own beds, too tired to begin to get to know each other.

That evening, Basanti emerged in a green and gold silk sari which clung to her figure closely. She was a lovely sight and I was overwhelmed by the very fact of her presence in the house. Till that time we had never spoken to each other, having had occasion for no more than a few shy smiles.

Where were we to begin? I still remember the expanding, tormenting moment of baffled wondering. I remember the sense of being lost in my ancestral home. I found myself in the presence of the only other inhabitant of a terrain that was all the stranger to me because I had assumed it to be my own native soil. What was one to do in such a predicament? Was I to tell her all I could remember, all there was about myself — or would some lesser and more formal credentials be acceptable?

Unexpectedly my confusion was resolved. I looked at Basanti. She stood before me, calm and poised, undisturbed by the dizzy eddying of my own feelings. Her look asseverated the fact of arrival in my territory, and proclaimed her expectancy that I would make it known to her — make it over to her. I could not but respond, and going up to her I took her hand.

'Basanti, this is your home. It has been waiting for you. Come, let it greet you!' I managed to keep my voice steady.

71

She smiled, her fingers twitching in mine, involuntarily responding to my invitation.

We went out to the garden. Immediately I sensed the lightness of her step. I glanced at her and saw a surge of animation rise to her face.

'It is so beautiful,' she said in a low musical voice, 'so young and fresh around the great old house.' They were her first words to me, and I remember the gladness that swept through me as I heard them.

I was excited when I attempted to respond. 'Now there will be beauty and youth within the house, too!' I was surprised at the nervous quaver in my voice. After a moment of shy perplexity Basanti turned to me, her face glowing with a pleased smile.

After that the words came easily to both of us as we went around the house. Then she wanted to see the view from the corner rooms on the topmost terrace. We entered the house and climbed the stairs. The wide terrace above the treetops and the sight of the gleaming city walls below us drew her out in joyful exclamations of delight.

We re-entered the house and I took her through the rooms, beginning from the top floor. She was both awed by and pleased with the vastness of her new territory. When we came to my mother's suite of rooms — which would henceforth be hers — she examined everything carefully, and her appraising glances and words came with assurance. Already she was the mistress in this sphere. I felt her grow in stature in this room, and that decided me. I had been hesitating as to whether I should, on the very first day, take her to the secret room. But now that I had observed her capacity to respond, I was curious to see whether she would react also to the sensuous beauty of the western room. I took her quickly through my father's suite, and leading her to the door of the secret room I opened it. The door to the balcony was open and the full light of the day was burnishing the rich furnishings. I saw Basanti stop for a moment on the threshold. She was startled. She raised her face to mine, her glance questioning me. Gently I led her into the room, saying nothing. Basanti walked across the room, straight to the balcony, as if she wanted to escape.

A light breeze blew a strand of hair across her face. She brushed it aside and said, 'It is so beautiful outside. Come let us go to the garden again!' I nodded and smiled at her.

We left the secret room. She had not been aware of it at all. She was young and unused. I felt very gentle towards her, I wanted her to open her senses, I wanted to help her, but wondered whether I would be equal to the task of assisting her.

That night after dinner she looked beseechingly at me several times, and then after some hesitation she said in a voice that begged for understanding, 'I am so tired. I know I should stay up and talk to you, but, tonight, please let me go upstairs and sleep.'

I went up to her and said that of course she must rest, adding, 'We will have many days and years for laughter and talk together. We need not rush ourselves, dear one.'

I watched her go up the stairs, and realized again that as a woman Basanti was innocent. What was I to do? Undoubtedly her mother had told her some things, but in the excitement of preparations for the marriage the meaning of her mother's words could easily have become blurred. She had remained innocent, and I admitted to myself that there was great beauty in the fusion of the freshness and the innocence of youth, both of which in her had remained intact. How should I go to her bed — not that night when she had asked to be left to rest, but the next night, or any other night — and subject her to the urgency of a sophisticated passion?

That night I went to my father's library and looked for

73

books on love and courtship. I found not only the well-known classics on the subject, such as the Kama Sutra, the Kok Shastra and some of the Puranas, but works by little known authors for whose writings my father must have sought diligently. Many of them were in hand-written manuscripts. I collected these works on one shelf, and I read late into the night. As I read on I learned that I myself was an unlettered novice in the intricate and vast culture of love. Before I opened these books I had thought that the main difficulties I would face in my relationship with Basanti would arise from her ignorance compared with my sophistication. I discovered now that both of us would have to be classified as barbarians in the art of love. Imagine how gauche I felt myself to be when I came upon a discussion of the significance of light effects in the erotic life. Form, colour, even smell — yes; but light! — it had never occurred to me that the gradations of light could vary the intensity and quality of erotic experience.

Taking from the works I read, and adding my own observations, I began to put down my own thoughts and feelings about love — I hoped it might help me in my new relationship.

The first fortnight of our marriage followed the pattern of the first day. By evening, Basanti would withdraw into her shell and beg to be allowed to retire early. I did not know what she did during the day to tire her, but by evening she looked so exhausted that I began to be alarmed for her. But then a change occurred. She would linger a while after dinner. Before retiring she would recline on the sofa by me. And on her face had appeared a slight flush which was deepening to a warm glow.

One evening we had dined early and were sitting on a wicker sofa. She had sat down closer to me than the size of the sofa necessitated. I took her hand and held it in my lap. For a moment her fingers were stiff, as if she were holding her breath. Then I felt her body relax and the warmth flowed to her hand. She let me draw her trembling body close. When I bent down to kiss her, again for a moment the warm flow seemed to freeze into a splintery tenseness. But that was the last barrier. We went together to her room. Her hesitation fell away and soon she was revelling in the warmth and pliancy of our embraces. It was a few nights, however, before we shared the fullness of bridal pleasures.

But it seemed impossible to break through Basanti's reticence in our daily life together. I would try to interest her in the tapestried room. She would come with me, but the richness of the room seemed to overwhelm her and she would move away to the balcony and look out at the vast western view. It was not that she was without emotional ardour, for often on the same night in the quiet snugness of her room she would provide for me a feast of tenderness and enfevered passion.

Then the next day she would be up early and I would meet her walking in the garden, a translucent pallor on her face and her eyes matching the quietness of the tentative early morning light. The previous night had stilled her passion and she seemed now to flow along intangible waves of communication with the little buds, the birds hopping from branch to branch, the rustle of the breeze, the spaciousness of the lawns and even with the distant cool fringe of hills. Thus she would become etherealized and so, too, would her awareness of me. It was only when I had come right up to her that she would notice me and greet me with a thin wisp of a smile. There was no sign of the continuous flow of tenderness of which the ancient books spoke.

It was in those early days of our marriage — about two months after Basanti had come to Amritsar — that Karam began to visit me again. I confess I found his return to our earlier habit of spending many hours together an obtuse intrusion. Fortunately, he understood the situation, and changed the brief visits in the mid-morning when I generally did wish to talk with someone while Basanti was in her private temple or reading quietly in her room.

Karam was like a pleased but insatiable animal. He told me so. 'Gyan, I am reaping a rich reward for my faith in marriage. It is much more wonderful than I thought. At first I did not even think my wife, Sita, was beautiful. But, of course, she is. One knows nothing till one is with a woman. She gives me endless pleasure, and it delights me to feel her ardent response to the pleasure she gets from being with me.' His eyes gleamed as he spoke, and his sensitive lips were smooth and full with unalloyed satisfaction.

I remembered that his wife Sita was a bony, plain-looking girl, but her shy eyes had a bright side glance and she had an engaging looseness of limb. I learned from Karam that the only cloud on his horizon was that soon Sita, who was pregnant, would go to her parents' home and he would be alone for several months before she returned with their baby. He did not know how he would face those solitary months to come.

I did not envy Karam. I regarded him as a relatively simple person who had fortunately found a like mate. His deep fires had only one source. Besides, my own life with Basanti was deepening. I had begun to feel and receive her waves of quiet communication, even in the early morning when they were at their most ethereal. Though there was still no continuing mutual flow of physical tenderness I did not despair. She was young, and both of us had much to learn.

Soon Basanti, too, was pregnant, and her coming motherhood brought a complex bloom. She gave herself with more tenderness than before, and yet she was more self-absorbed. I accepted this as an intricate self-balancing process in her life.

I suppose a similar process explained my renewed diligence in overseeing my business affairs. For the first time in my life I made a calculation of how much I had saved and how rapidly I could add to my capital. I found my financial position to be much stronger than I had imagined, and decided to buy up a timber concern. It seemed to me that there would be a sure and increasing demand for timber by the building and furniture trades and by the railways. After taking this decision, I wondered whether I was not being unnecessarily acquisitive. But I allowed the purchase to go through as planned, telling myself that it was reasonable that I should make some additional provision for the infant who would soon be born. My friends and acquaintances, including Karam, commended me for my wisdom, and I admit their approval was flattering — for this was my first venture into business. As it happened my assessment of the lumber market turned out to be more or less correct. The business venture expanded rapidly, earning large dividends.

Basanti's parents had taken a house in Jullundhur, and when there was still several months to the birth of her child she asked if she could join them, for her parents had taken the house for her convenience.

'But of course you can, unless you would rather wait here till the event is somewhat nearer,' I replied, searching her face for her own inclination. I did not know whether she was going because she wanted to get away or because she wanted to please her parents.

The girl looked confused and troubled. Her face flushed and she said, 'Of course I would like to stay with you. This is where I belong,' and she looked up at me to let me see that her eyes confirmed her words. Then she

77

went on, 'But I might not know what to do here if any-thing should happen.' Again she looked at me, expressing her fears and betraying an undertone of lack of confidence in me.

I felt I should not attempt to put into words what her eyes revealed; besides, I might be mistaken; and after all, how could she have confidence in me in a situation with which neither of us was familiar? We agreed that she should leave as soon as she could get ready.

When she left, to my surprise I did not feel I had been left alone. I was more constantly aware of her than when she had been in the house. I even found myself wandering into her rooms and, what I had not done so far, I even went and stood at the entrance to her little temple. The room was attractively decorated, and I thought some of the drapings were familiar. I went to the secret room which I had not visited for weeks because Basanti seemed to shrink from using it. She had taken some of the silk drapings and bolsters to the temple. I smiled, wondering whether she realized what this room must have meant to my father and his wife. Whether she did or not, I was pleased that one aspect of the beauty of the room should have affected her sufficiently to have led her to remove it to another context.

I again took to practising swordsmanship early each morning. After an early half-hour at sunrise of this sport I would stroll a little in the garden, and I was amazed at my capacity, without developing it and without con-sciously exerting myself towards it, to project myself into the life of another person. It was as if Basanti, rather than I, were walking in the garden. I felt a vivid sense of communication on a complex wave system with the life of the garden, with the high flying of the birds, and with the line of the Himalayas against the northern horizon. If someone chanced to greet me, I would look past him as Basanti often had looked past me during her early morning stroll in the garden.

This experience of fusion with Basanti's personality and the unremitting restlessness of the nights spent alone began to trouble me. Was I over-immersed in her? Was love so enveloping that in it one ceased to be? Should I extricate myself from my plight of the devoted lover — and could I? To ask this dual question and to struggle to withdraw, at least mentally, from my state was to discover again the power of love. Its completeness held me. Helplessly I would cease struggling, telling myself that it would be folly after all to seek again the despair of being isolated, which would lead once more to the search for brief moments of release in work, or contemplative indolence, or social or political rivalry with my fellow men. Whichever way I argued, in effect for most of the time I lived deeply immersed in my love for Basanti.

It was during this period that I read a brief notice in a Calcutta newspaper, published in English, to the effect that Jennifer Trip had been thrown from her horse while pigsticking in Central India. She had been rushed to a hospital but had died before medical aid could reach her. I read it again. Thrown from her horse and killed! I felt myself sinking into despair.

I was in this state, immersed in both love and death, when news came of the birth of a son. Basanti wanted to name him Man Mohan — the heart and mind of love. I agreed, for her suggestion seemed to come from the very depths in which I was now living. I went to Jullundhur. My appearance startled Basanti. I had lost weight and looked troubled, she said. But she herself was radiant with joy, feeding her infant and then watching him fall asleep.

I returned to Amritsar and had the house repainted in preparation for the return of mother and son. It was in

79

the same spirit of preparation that I turned with renewed attention to my business affairs.

Karamjit's wife, Sita, had returned from her parents' home with a strapping baby boy, and from then on I saw little of Karam for many months. For the last two months before his wife's return he had become quite impossible. His slow turbid nature had stirred itself up to the seething point. I could almost smell the hot unpleasant humour into which he had fallen; and on any subject of conversation he had become poisonously argumentative. His usually rather deep voice would rise to a monotonous treble and he would argue as if his very life depended upon it.

I was very glad when his wife returned. He delighted in her big boniness; and she was undoubtedly very good for him, for sometimes they visited me briefly with their young son and Karam was again his beaming, satisfied, animal self.

I had many other acquaintances in Amritsar, but I was so engrossed in my own life that I cultivated none of them. Narinjan Das, however, was thrust upon me. He had been a general clerk at the timber concern when I acquired it. The other white collar workers of the firm had been relatives of the previous owners and they had gone elsewhere when the ownership changed. Narinjan Das probably expected that he, too, would have to find some other employment.

On my first visit to the timber yards he had followed me about in great agitation, and had acted as a sort of unsolicited guide. At that time I did not even know who he was. I did not inquire, but in between information about the yards he would sandwich personal intelligence so that his remarks were most disconcerting. His words came in the subdued yet emotional pitch that is often emitted by one who is both most anxious and yet fearful that his efforts are failing in their purpose.

'Sir, this — as your good self sees — is heavy deodar timber from Chamba. It arrived last year. That was when, through misfortune, my salary was reduced from forty-five rupees a month to forty rupees a month.'

I wished the man would leave me alone; and yet I wondered what all this was about. 'But why did the arrival of the timber lead to a decrease in your pay?' I asked.

'Honoured sir, it was entirely due to the unkindness of the stars. For me, Jupiter is a very evil star, and at that time he was so fiercely in the ascendant that nothing else was possible.'

Madcap, I thought, but I was silent. I knew that if I burst out at this absurd-sounding statement, it would hurt the nervous obsequious man standing respectfully three paces behind me. Indeed, he must have been hurt at my silence, for within a few minutes he sought an opportunity to enlighten me further about his fortunes.

'Your Honour is now looking at the sawing machinery which the late owners bought, but as your discerning eyes must have noticed it has rusted unused. Money wasted! and so it had to be, for they dismissed workmen to collect the funds to buy the machinery. How then could the venture prosper, honourable sir? The machinery was never put to work; and there was double expenditure because men had to be re-employed to saw the timber. That was the time when my dutiful wife, perplexed by this treacherous world, bore me another daughter instead of a son, thus creating a stupendous problem for a poor family which already had a daughter!'

At the end of the tour he bent down to touch my feet. Hastily I moved away as I said with spontaneous severity, 'If you have anything else to say, please stand up and say it, my good man!'

Had I hurt his feelings? No! Courage and joy filled

his round, not unpleasant face. Under the white turban the wrinkles on his wide and protruding brow were immediately rubbed off. His small eyes, however, were blinking so rapidly that I found it difficult to say what they looked like. He wore a clipped moustache over his small mouth — the mouth of a person who has persuaded himself that he has been given a small orifice because it is his destiny to say small things. His shortish figure was clad in a very shiny blue coat, a colourless striped shirt, and bedraggled white loose trousers.

He stood up straight, rubbing his hands together and saying, 'Protector of the poor. Kindly do not let me forsake this business. I am a poor man, living under evil stars. I will be willing, honourable sir, even to accept a cut of another five rupees. Pay me thirty-five rupees per mensem but do not order me to depart from your honourable presence!' He raised his folded hands in supplication and his eyes were wet with tears.

The managers of my landed properties were shrewd men, generally ingratiatingly polite, and they knew, too, when to be sullen or devious. But I had never been confronted with the sort of behaviour that I had just witnessed. I asked the man his name and what he did. He replied in his strange manner. I wanted to tell him at once that he could have fifty rupees a month instead of forty, and that he could count on staying on as long as he worked reasonably well. Even fifty rupees a month sounded to me a pitiful income for a man with a wife and two children. But I had no knowledge then of the circumstances of the concern which I had just acquired, so I said to him:

'My good man, I will not cut your pay. Continue to do your work and I hope your lot will improve.'

Again he wanted to fall at my feet, but he saw my forbidding look and he desisted. 'Great sir, in spite of

Jupiter you do this for me! Your good stars must be very powerful. God will bless you, noble sir!'

I nodded and walked away quickly, not knowing what to make of the man and determined not to let him embark again on his abject talk about the stars and himself. Narinjan Das worked for me with unbelievable devotion. He was required to be at work at the timber yards at nine-thirty each morning. But at about eight-thirty each day he arrived at my house on a very rickety bicycle, and remained there in attendance till about nine-fifteen, giving me — if he caught me alone and I did not appear to be preoccupied — special intelligence about the timber market which he had picked up in the town. Sometimes he would add other information, such as the news that he had been to all the wholesale fruit merchants and discovered that the best pomegranates in the locality came from the village of Jallo, adding that, as pomegranates were very good for young mothers, perhaps I would like to send a man to buy up the yield of several trees for my wife. 'Moreover,' he would continue, his round face beaming as if this were the most important discovery of the day, 'the stars are very favourable for such a venture at the present time!' Because of his abjectness and his goodness of heart I never broke out and rebuked him for his astrological talk, although whenever I thought about Narinjan Das, I promised myself to chide him about it. But I could never quite say this to him when he appeared before me, for, as I observed him, poor Das did behave monstrously like a marionette who was being motivated by some distant creator — perhaps the stars, after all!

When Basanti returned with our son the old house came alive. The return of mother and son brightened the whole scene — the gardeners wanted the garden to look its best, the servants scrubbed the floors and furniture, and even the rather wily managers wore almost pleasant looks. And Narinjan Das and his companions at the yard brought garlands and showed genuine delight.

It was as if a new dimension had been added to our lives: and I suppose, in a sense, that is what had happened. The child projected us into time, making me aware of the endless continuity of life — there would be Man Mohan's children, and then their children and so on, going on alongside time, and somewhere in the far-off future perhaps transcending it.

But much more than all else it was my feelings for Basanti that overwhelmed me. I had no doubt now that I was deeply in love with her. She opened up in me the tenderness of affection and the burning urgency of passion. I could not take my eyes off her; and my hands, my chest, and my whole body became as if flayed to rawness with the fever of the unslaked longing to touch her. I must have been gauche in my advances — I told myself there could be no other reason for her shy and fleeting responses. I was distraught, but tried to reassure myself into a calmer frame of being. She had been away for many months and we were still relatively strangers, so that her shyness was to be expected. A little while and she would warm towards me.

It was like beginning again our married life together. Only, to the difficulties that had attended upon our first effort was added a diversionary force. Although there was a host of servants, and a trained nurse for Man Mohan, Basanti was engrossed in the new young life. At times I was concerned, for her devotion to the child was wearing her out. Man Mohan was thriving wonderfully. I had been surprised at his soft light-brown hair, but as he pulled at his mother's breasts and sucked at his supplementary foods, his hair darkened and his large face grew purposeful — intent on more and yet more nourishment. For hours together he would wave his plump arms and legs, desperately trying to grab at life. It was not till many weeks had elapsed after her return that I realized that Basanti's shyness arose not from a sense of being a stranger but from her engrossment with Man Mohan and from her uneasy effort to conceal the fact that, at the moment, she had no great interest in me. Perhaps that was not quite fair to her. Perhaps she expected me also to be engrossed in Man Mohan and to meet her on the basis of a mutual parental preoccupation.

I suppose the clamour of my being for a direct flow of tenderness between us blinded me to the complex path which lay before me. I wanted urgently to give her my tenderness, and felt rebuffed and perplexed by her cool, gentle disinterest. I tried to make myself clear to her by gestures and by words that hinted at the need I felt for a closeness with her, but her beautiful shy face would turn away to attend to some chore for little Man Mohan. I felt completely unused. Was I, perhaps, too ardent? I began again to study the ancient books on the married state and learning from them; my imagination was filled with delightful pictures of tender bliss. When Basanti would let me be with her, I met her with these glowing prospects filling my mind. But the circumstances were never propitious: time would be short, she would have other things to do, and the glowing colours of my pictures would disintegrate into a memory of the cold print from which they had been conjured.

Still, life could surprise me. Came the evenings when Basanti was at a high pitch of tenderness and passion. She whom I had come to regard as unchangeably silent and shy became a stream of hotly breathed utterances that stormed over me. Wonderful hours would follow and I would see the beautiful pictures of love which I had

85

yearned to bring to reality becoming our real life. But the flutter of passion would pass, leaving a painful limpidity in which the more stable facts obtruded with hard clear-cut forcefulness: my frustration, Basanti's absorption in the young child, and her acceptance of life as it was rather than as an adventure to be shared with me.

Basanti began to visit the library, and took away books to read. I was pleased that she should, and between us there developed a stimulating companionship through our reading, which did something to dissipate the dull feeling of frustration that I had on a more personal account. Often the absence of a continuity of physical tenderness was an unbearable emptiness, but unexpectedly I would find relief in some beautiful poem or in a pleasant discussion with Basanti about a new book. And when even my books failed me there was the booming sound of the logs being stacked in my timber yards.

Basanti was in such a quietly engrossed mood that it was not until Man Mohan was almost three years old that she conceived again. She went to Jullundhur as before, and our second son, Romesh, was born just after the end of the First World War. She returned home even more engrossed in her life as a mother. Indeed, so strange is motherhood that she seemed deliberately to improvise severity towards Man Mohan to whet the edge of her love for Romesh. She would say that it was because Romesh was a much more delicate child than Man Mohan had been; that she had to give him all her care and attention; that Man Mohan did not understand this and insisted on demanding all her attention for himself. This argument was sophistical as Romesh was at least as sturdy a baby as Man Mohan had been. But there was no arguing with Basanti. I tried often enough. I appealed to her, but nothing had a worse effect than my saying, 'Do be

86

reasonable'. For days after, she would barely speak to me, as though I had deeply wounded and insulted her.

Is one never to criticize when one loves? Is understanding between people to be so complete that criticism becomes superfluous, pointless? At that time these questions troubled me. I think I know the answer now. But then I was disturbed by the devastating effect that my criticism would have on Basanti. And I had to admit that she never criticized me.

During the war years there was a brisk demand for timber, and I found myself expanding my interests and working in terms of very large sums of money. I decided that, as soon as the war was over and I could import the requisite machines, I would set up a wood pulp and paper factory.

It was Narinjan Das, odd mixture though he was of astrology and submissiveness, who among all my employees responded most eagerly to the expansion of the timber business. The others were glad to see the business grow, and as my rising profits enabled me to increase their wages they had reason to be satisfied. But Narinjan Das was constantly pointing out new avenues of demand for timber, and on his own initiative he sent a man into the Chamba hills to locate unexploited forests. I was then able to persuade the Forestry Department to frame silvicultural plans for some of these tracts which brought us an increased supply of logs. I warmly commended Narinjan Das many times for his enterprise, and invariably the only response I received was, 'It is a deep honour, sir, to repay in small part the debt of gratitude I owe to you who have rescued me from the influence of Jupiter.' He said this with deep emotion, so that I had to suppress the smile that was rising to my lips.

'Well, Narinjan Das, I want you to take over the

management of this timber yard. I know you will run it well, encouraging and helping the others, and seeing to it that they are well treated. Do not hesitate to tell me if you think anything should be done for the men.'

Narinjan Das looked confused. His usually ruddy cheeks turned pale. He shook his head, and clasping his hands he said in a low, quavering voice, 'Is Your Honour wishing this upon me for the future, or is Your Honour commanding me now to this great and noble duty?'

I put my hand on his shoulder. 'Narinjan Das, I am appointing you manager now, because I know you can take charge today of the yard. I have confidence in you.' I said this quietly and slowly so that the full import of my words should sink into his startled brain.

The colour returned to his face. His eyes, for the first time that I could remember, became calm and purposeful. He drew himself up straight, and then with an even greater submissiveness than he had ever used, he said, 'Protector of the Poor ... great sir, I will fulfil your command and you will see that Narinjan Das will bring greater honour still to your illustrious name. May God bless you richly!'

When he took over as manager, Narinjan Das bought himself a new navy blue serge coat and a white silk turban. He worked even harder than before and lost enough of his awe of me to permit himself to smile when he greeted me. Then about a month after his promotion, he begged my wife and me to honour his home with a visit. He told me that he had just moved to better quarters and felt that he could receive us in a fitting manner.

We drove in our open carriage to an old part of the city, and stopped before a plain old wooden doorway set in a crumbling outer wall. Narinjan Das opened the door and took us across a small courtyard. We climbed the narrow dark stairs to the second floor veranda of the narrow house in front of us. We were greeted by a woman with a sunken chest and a pale yellow face, looking uncomfortable and somewhat bewildered in her heavy silk sari and traditional gold jewellery. It was Narinjan Das's wife. Behind her stood two girls, one of them about seven and the other barely able to stand. They were neatly dressed and greeted us with bright *namastés*. On the tiny veranda were three old cane chairs and a *charpoy*. On these had been spread beautifully worked *phoolkaris*. We were seated in two of the chairs. Narinjan took the third, sitting very straight on the edge, and his children climbed on the *charpoy*. His wife went into the house and returned with an array of sweets and highly scented sherbets. Narinjan Das jumped up and helped her to serve us.

His wife broke the awkward silence, speaking almost inaudibly. 'You have honoured us, Rai Sahib, and Your Ladyship. Our humble home is at your disposal from now on. These children will do as you bid them.' And the little girls listened unflinchingly, like little heroines preparing for the worst. His wife's announcement made, Narinjan Das smiled proudly. This was the climax of his purpose, and volubly he repeated and embroidered her sentiments.

There was little other talk, but all through the visit I was deeply moved by the dignity and strength of Narinjan Das's family. Though the mother and the children clearly had been underfed and underprivileged in every sense, no bitterness seemed to afflict them. They insisted that we enter their two small rooms. Both were spotlessly clean, with shining pots and pans, neatly made beds, and a little library of Hindi literature. When we came to the shelf of books, again Narinjan Das's face beamed as he told us that the older girl went to school, while the mother read stories to the younger. An extraordinary tenacity of tradition had preserved some grace and beauty of life for these people through all the long miseries of near starvation and ill health, and the suffocation of the one room in which they had lived for ten years before they had moved to this new and more spacious home.

As we were leaving, Narinjan Das's wife had an uncontrollable fit of coughing. She sat on the *charpoy* and the cough came deeper and deeper from her sunken chest. Narinjan Das held her head and kept saying, 'It is nothing. It is nothing.'

But she started to moan between her fits of coughing. Then, in silence, she half collapsed and half deliberately laid herself down. She lay silently, her children looking at her wide-eyed, the elder girl gently smoothing a gay jacquered cotton blanket over her.

The fit of coughing was over. We stood silently for a few minutes, and then, with Narinjan Das, we went down the stairs.

'Narinjan Das, I will send you our doctor,' I said to the troubled submissive man at our side.

'Sir, do not send your doctor. It will be like our reaching for golden apples. Let me send for the doctor in the next street. He will take only a small fee. God knows if he has any skill, but I can pay him and he will do his best. If it is in her stars to recover all will be well. Jupiter is again in the ascendant these days.'

It would not have done to send our doctor or to offer to pay for him. Narinjan Das, submissive and obedient though he was, had too much self-respect to accept charity. Fortunately, Basanti had another idea. An American woman doctor had arrived in Amritsar, a Dr Hauser. She was attached to the mission hospital. We arranged for her to visit Narinjan Das's home. It turned out to be a case of under-nourishment and general debility. Fortunately Dr Hauser was able to give the patient something more than the medical care and treatment she required. Her own warmth and quiet efficiency meant more to the poor woman than the medicines. A few months later Narinjan Das mentioned to me that he was again taking his wife to Dr Hauser. I was alarmed, but Narinjan Das smiled, and blushing as if he were a young bridegroom, he said, 'Honourable sir, may God grant me a son. My wife is again to bear a child to me.'

But it was again the season of Jupiter, and in due time another daughter was added to the Narinjan Das household.

For me, the urge, fury, and sweetness of life more and more centred around Basanti. Life with her was the most complex experience I had ever known. In part this was due to my own phase of existence at the time. My desire to make amends to my father had been frustrated by his death, and I think I was left with a dominating sense of urgency to make life good for Basanti. I did not realize that in permitting this urge to grip me, I was lessening the chances of developing a satisfying relationship and was reducing myself to the position of a servitor. Not that she would have agreed that this was so, nor was there any suggestion that I was embarrassing myself. On the contrary, she gave me all the respect that a woman brought up in our traditions gives her husband. But there were indications that she was aware of the difference in our roles created by my anxiety to do all I could for her. It was this awareness which would lead her, in spite of her retiring and phlegmatic nature, to behave with extraordinary wilfulness.

I could never predict what Basanti would do. Not that she played at melodrama. Hers was a deep, still current which made it all the more difficult to sound. One whole winter she saw only her religious teacher. She ate in her own rooms, and very early in the morning she would walk in the garden, leaving the trail of her sandalled feet on the heavy dew and in the hoar-frost. The trails and the tinkling of the bells in her temple were all that I ever saw and heard of her that winter. She had sent her younger son to Jullundhur to be with my sister, and the older one — Man Mohan — was at school in Delhi.

Without warning she emerged from her winter of meditation. She did not say a word about it to me: neither why she had been led into it nor what she had gained from it. What was equally unexpected, though welcome, was that a few days after that phase of her life was over it became clear that her experience had not made a recluse of her or added melancholy to the natural reserve of her temperament. Instead, she responded gaily and strongly to life around her. She asked about the new books that had come in during the winter, and avidly read many of them. She wanted to take up riding again and I bought her a beautiful riding pony. What is more, Basanti was eager to renew our intimate life together. She gave herself with impetuosity, and there was every indication that it was increasingly pleasurable to her. But, equally, she withdrew farther into herself on each occasion — I suppose in a sense she came to me from farther away than before: her personality had become more spacious. Her attitude provoked me to greater ardour, and I redoubled my researches in the literature on the subject. In a technical sense she let me use the results of my knowledge, indeed she enjoyed my doing so, but almost before the experience could spread out and become a glowing thing between us, she was gone — having retreated far away into another part of her life.

Towards the children her behaviour was at once a model of equanimity and a range of expressiveness that must have baffled them. She never lost her temper with them, she was frequently playful, very frequently reserved and un-approachable, and often their guide and friend. The most apt description of her behaviour is to say that she was regal with them. Perhaps all mothers are also queens. Basanti certainly was.

Though she came out with me on some social calls, or even at times accompanied me to the yard or to visit Narinjan Das's family, she was always remote on these occasions. I do not think she ever could have told the difference between a log and sawed timber; and she barely remembered the names of our Amritsar acquaintances. She never invited them to the house.

The Narinjan Dases invited themselves. That, in their case, was merely a development of Narinjan Das's earlier habit of coming alone early each morning. Often he would do that too, but in addition — after asking my permission — he would, of an early evening, bring his wife and their three daughters to call. It was an act of presenting themselves before us, of being available, or of displaying loyalty. My wife occasionally acknowledged their presence by entering the first floor drawing-room for a brief time while they were there, but generally she absented herself. They expected me to spend a half-hour with them while the girls were served with sweets, soft drinks, and perhaps given a toy or book each.

I tried gently to persuade Narinjan Das to let me defray the cost of the education of his children. His eldest daughter was at the local college, and the other two were in school. But his response the next morning was to bring my wife a beautiful *phoolkari* which his daughters had brought back from a visit to relatives who lived in a neighbouring village. He looked very pleased as he opened the beautiful material and said, 'It is worthy of Her Ladyship, honourable sir. It is embroidered by the most famous seamstress in all the Punjab. And sometimes

93

when the stars are favourable her work excels, as in this piece. My daughters are wise girls to have brought it for the Rani Sahiba. Such wise girls will look after themselves well. I have nothing to be anxious about.'

In spite of his subtle reply to my suggestion, I was unhappy at not being able to help him. However, the situation solved itself in another way. In spite of the depression and other difficulties, my business affairs were doing very well. Our pulp and paper mill was the first of its kind in North India, and in it the timber yards had a steady customer. I felt I should acknowledge Narinjan Das's loyal work in a way which would give increased scope to his deep interest in the yards. I decided to give him a share in the profits. Again he was incredulous when I told him of my decision; but each quarter I insisted on the sharing of profits, and to oppose me directly was not in his nature. I felt relieved on his account. In time, now, he would be comfortably off and his daughters would not be without the elaborate trousseaus and jewellery required of most young brides. It turned out to be a timely gesture on my part, for a few months later Narinjan Das's wife contracted pneumonia, and in spite of the efforts of Dr Hauser she did not live through it. Narinjan Das was demented, and terribly upset for his daughters. He did not say so, but I knew that the fact that at least he had no financial worry on their score did help him.

Basanti had come rather to resent the frequent visits of the Narinjan Das household. Now that the mother was dead the daughters appeared àt our house very infrequently. Narinjan Das said they were busy with their studies; but they were also normally sensitive young girls, and it must have been obvious to them that they were not exactly welcome to the mistress of the household.

For many years Basanti had not conceived. As she

94

approached thirty-five she wanted to be a mother again. She asked whether we might call in Dr Hauser to examine her. For many mornings Dr Hauser was a familiar figure at our house in her spotless white organdie dresses, her heavy horn-rimmed spectacles, and her mass of golden hair over her smiling face and grey-blue eyes. Basanti particularly liked her, and, surprisingly for so aloof a person, wanted her as a friend. After a month's treatment all was well, according to Dr Hauser.

Soon after, Basanti conceived again. As this had been her own wish I expected her to be pleased. But the change that came over her was much greater than just a deep feeling of pleasure. She became suddenly aware of the house itself and of everything around it. I would come across her admiring the frieze on the ground floor walls. It was beautiful work done by the master masons of the eighteenth century, but before this sudden awakening Basanti had taken it for granted. Always she had loved the old balcony outside the secret room, but thus far she had never entered the room except to ransack it for drapings. Now she was often in it and admired the Rajput paintings so much that she begged me to be on the look-out for more fine examples which she wished to hang in her rooms.

I was overwhelmed and baffled by the change. There were moments when, though I am not superstitious, I was even filled with misgivings. It was a wonderful flowering of her personality, but the question kept coming to my mind — why now after all these years? I was very tender with her, and in response her face and eyes would glow as they had never before. Whenever I could shake free from my inner questioning it was like a new marriage, or, rather, it was a marriage at last. I ceased being her master servitor; spontaneously and constantly our thoughts and our desires would meet, and for no reason Basanti

95

became gently solicitous about me. I felt a thousand dams within me being washed away and a great torrent of warmth and life coursed through me.

Delightedly I bought for her a beautiful series of Basholi Rajput paintings — in brave oranges, blues, green, and ochre. They told a story of the arranging of a marriage, the coming of love, the dutiful wife cooking for her husband, the confinement of the mother, and then sustaining love depicted idyllically against a background of flowers, birds, and the whole world renewing itself. Basanti hung them in her bedroom where she could see them when she awoke in the morning.

A crisp joyfulness lit up the household. Our sons, Man Mohan and Romesh, were approaching manhood, and they responded to the new mood 'of their mother by completing their own growth out of the remnants of childhood which linger when children are brought up entirely at home. Their mother's warmth and flowering seemed to transmit courage and shape to their personalities. Man Mohan suddenly stood before me as a man, and asked to go and work in the timber yards. Romesh talked of going abroad. Both of them would invite their friends to the house freely, enlivening our outdoor dinners in the late spring of that year. I felt my own life at last attaining the deep quality of joy and sparkle, as well as the gentleness and warmth, for which it had craved. In my library I sometimes opened the old books on love but I did not need to read them now. I smiled and patted their covers: I knew their contents.

When the time of Basanti's confinement approached, Dr Hauser — who had been looking after her — suggested that as it had been many years since the birth of her last child and she was no longer young, it would be wiser for Basanti to be confined at the hospital. Basanti readily agreed, and though I was not happy with the idea I raised no objection. Dr Hauser reassured us. All was going well, and a few days after the event both mother and child would be back at home.

When her labour began, Basanti was rushed to the hospital. The pains became very severe and Basanti seemed exhausted. But Dr Hauser was not alarmed, and Basanti, always very brave, never uttered a cry. She was straining towards the final labour when she went into a convulsion and lost consciousness. Dr Hauser was confident it would be easy to revive her, and sent word to me that soon all would be well. I remembered Dr Hauser's competent kindly expression, and my twinge of apprehension disappeared. I was expecting the news of the birth of our third child, wondering how it would be for this new little person to grow up with elderly parents, when I was asked to go immediately to the room outside the maternity theatre. Dr Hauser came to me, her usually cheerful face red and swollen to almost unrecognizable shapelessness. Basanti had never regained consciousness. The strain of her labour had been more than her heart could stand. She was dead. And they had not been able to save the infant life within her.

Life had been snatched away at the threshold. Basanti and I had arrived there only within the last few months. At last our marriage had begun, and the earlier years had seemed like a long arduous courtship — during which I never knew whether she would accept me eventually. But it had been the threshold of illusion and death, leaving me gazing into an endless waste of darkness.

Social customs are an orderly filter through which we pass the rude facts of life. Generally I prefer to do without the filter, but on that occasion, though I wanted to be alone, later I realized how I was helped by the social response to my situation. In a few hours the house was full of people: Basanti's parents, my sister from Jullundhur, even my old mother from Ajnala, Karamjit and his wife, Narinjan Das and his daughters, the Municipal Commissioners of Amritsar, most of the lawyers of the town, all my employees, farmers from my lands; and, perhaps drawn by the great crowd, almost all Amritsar seemed to collect or call during the next two or three days. At first I was exasperated by their presence. I wanted only to be alone; I did not want to lift my face to any other human being. I was crushed by my sorrow, but I had to steel myself to the effort of meeting these hundreds and thousands of people. They were so filled with grief on my account that I felt them taking part of my sorrow, and that carried me through those days. How long that unbroken spell of blinding sorrow lasted I do not remember. After many days, I became conscious of recognizable individuals around me: till then they had all been part of the vast number of people sustaining me. Coming to life again was literally a matter of learning again to see, to walk, to hold up my head, and to close my mouth to the involuntary sighs and groans that filled me.

My sister wanted me to go with her to Jullundhur, but wisely she did not insist. It was better for me to stay at Amritsar and to accept the fact that the world went on as before. By letting that fact intrude on the enclosed solid darkness of my grief a blending took place, and the fact of Basanti, our life together, the sudden flowering at the end, and her death — all became part of the world that went on.

Some months later, I was able to make a trip to the hills. The mountains lifted themselves up in a great splendour of grey, green, and white. The clean pine-scented air, the calling of the shepherds, the brisk cold of the morning, the limpid air of the twilight filled with the long swinging calls of the cicadas, then night falling quietly or being brushed up from the depths of the valley on the winds ruffling the fir trees; all these things seemed to weave me together again, physically to reintegrate me and spiritually to give me balance and peace.

I felt I could face people again. I went to Delhi to visit Man Mohan who was in his final year at college. He met me, full of uncertainty, not knowing whether he could carry his own load of sorrow, let alone stand unflinchingly in the presence of our stupendous joint burden. But when he saw that I was strong, that my anguish was part of my life and that I was the richer for it, he too regained a sense of assurance. Then we could talk of Basanti as a reality in our lives. There was no loss now, only the pain of our too intense realization of her, and the sad gentleness that flowed from us to others came back as a salve.

Part IV

CALLING on the Karamjits, which I was doing, was still an unfamiliar experience. It was like putting on a new suit of clothes for the first time — even though it has been tailored its lines follow their own law till the material yields to the curves and hollows of one's body and limbs; then gradually it sheds also the smell of the loom. Through my youth, and again on my return from England, and during my married life, I rarely went out in Amritsar: it was our home that had always been the accepted meeting place.

So now, though I had over the years visited the Karamjits three or four times, I felt acutely out of place. Karam did not help things by remarking in his usual blundering manner, 'Gyan, come, you are still so ill at ease in our home! When will you accept it as your own? Do relax, my friend.' His comment was so true that I reacted defensively — just as one would defend one's new suit against the criticism that it was ill-fitting.

'Nonsense, Karam. I am at home here. I can't help it if I am the sort of person who doesn't look the way he feels.' It was not an impolite reply, but it was pure dissimulation and added to the irritation caused by Karam's hospitable injunction.

Fortunately his wife Sita entered the room, and Karam dropped the subject. Besides, just then there was the full-throttled snorting of a car being driven at high speed up the driveway. A moment later there was the smart bang-to of the door of a roadster, followed by the entrance of a genial-looking man whom I had not seen before. His small greyish eyes twinkled in a round bearded face. He had a friendly looking small straight nose and a mouth which promised to be both kindly and teasing. He wore a British style sports jacket of Harris tweed, grey flannel trousers, and the light suede brogues that were part of the usual attire of the sportsman. He quickly folded his hands to Sita in greeting, and then, opening his arms, he went up to Karam and embraced him. With a rippling smile beading his eyes and displaying his short well-formed teeth, he was saying in a throaty, sensual voice:

'Going well, going well, young cock? Tell me later, my boy!' And he glanced meaningfully at Sita.

Then he came up to me, holding out his hand, while Karam introduced him. 'Meet Ranjit — remember his father, D. J.?'

Ranjit gave me a warm handclasp, and if I still looked out of place it did not put him off. Indeed, from the moment he entered the room — though judging from his appearance the type of life he lived was farther removed from mine than was that of Karam and Sita — my uneasiness vanished. Immediately I felt myself in easy contact with him.

'Well, it's good to be back in old Amritsar! You know,' he said, his eyes twinkling at the stolid but smiling and pleasurably interested Karam, 'I have been away for a month in Rampur, Bhopal and Bharatpur. Wonderful part of the country — such good pigsticking and shooting, and best of all ... oh! I can't tell you just yet.' He looked at Sita, managing to appear decently shocked at himself as he went on in his pleasant way, 'I am too disreputable a fellow to talk in your presence, Sitaji. You must forgive me.'

With a forced smile, and a quick toss of her head, Sita responded, 'In any case, I was about to return to my household duties. And I am sure Karam also wants to tell you things which I had best not hear!' She got up and walked out, throwing a pout and a clouded look at her husband, and ignoring his protestations.

Karam shook his head. 'Strange folk, women. Must always be on top, even when they are beating a retreat!'

'We must let them feel they are on top! They deserve at least that from us,' chimed in Ranjit with the tone of 'the man who really knew', adding 'we are a selfish lot, we men.'

'Well, anyway, out with whatever it was that you had to reserve for our masculine ears,' said Karam, a flush rising to his face.

'But how can I tell it? It was beyond all that I had ever known.... They were wonderful. Their eyes black as a doe's, their heart-throbs quick and hot like a hunting leopard's, and their embraces like a vine creeper. What hours and hours of bliss, my friends! How fortunate I have been!'

Karam had opened a bottle of *thurra*, a fiery North Indian spirit, and poured a stiff peg for Ranjit, and a small one for himself and for me. Handing Ranjit his drink, he said, 'So here we thought you were living a clean manly life — shooting, riding, and so on, and I must say you look tanned and trim, as though you had been — but all you have been doing is plain and simple carousing, Ranjit, which you could have done here in Amritsar. Anyway, drink up, young cock!'

Ranjit took a gulp, and as the fire coursed down his gullet he exclaimed, attempting to shrug off the pleasurable sensation, 'Ah! What's the good of this stuff alone! ... But what's all this about my not living a clean manly life. Don't I look well? — you said I did! Nothing more manly than that utterly divine sport, my friends. Look at me — muscles all bulging, cheeks toasted, and eyes clear. I can tell you, Karamjit, there is nothing plain and simple about carousing as you call it. It's a man's heaven. Say what you like, that's what it is. I am sure your friend Gyan would agree.' He added the last few words giving me a look over his glass as he took another gulp of *thurra*.

There was an infectious warmth and sincerity about the way he spoke. He exuded the pleasures he was recounting. I must have looked as if I were sharing in them, for when I smiled and said rather weakly in reply, 'Yes, that is how it must be', he stood up and coming towards me, he announced, 'We are going to like each other, Gyan. We understand each other. As for our friend, Karam, we will teach him a few things. Maybe he will learn after all. Karam, don't stand there blushing like a girl!' He had turned to face Karam.

I was glad Ranjit had directed his last remark to Karam. It gave me a moment to reflect on what he had said. In a sense, I was taken aback at the statement that he and I understood each other, and yet it was true — even if I hardly knew what he was talking about. Besides, it flattered me that he thought I understood him better than did Karam.

But Karam was not accepting the lower category to which he had been assigned. To my surprise it was he — not Ranjit — who was eager to put to the test his sophistication, which had been called in question.

'Blushing, indeed! My dear chap, I could arrange such a wonderful time for you here in Amritsar that you would forget for ever this insignificant visit of yours to Bhopal and Bharatpur! Would you like me to?' Karam was pouring himself another drink, and the flush on his face had deepened.

'Wah! my lad. Wah!' said Ranjit, good-naturedly. 'We saw a few minutes ago how much you could do in this house. But I understand. We are all respectable people and you are a married man. Don't take on any more responsibilities, my dear Karam. You have enough! Leave these arrangements to Gyan and myself — we will manage things for you!'

Again I had been flattered, but I was now being led into a situation beyond my depth. How was I to help with these arrangements? But the situation mesmerized me and swept me along so that I found myself supporting the point made by Ranjit.

'Yes, of course. My house is at your disposal. We can arrange some good entertainment there.' I stopped. Had I let myself in for too much? — I was not deeply enough under the spell of the moment not to be assailed with doubts. I covered up my tracks, adding, 'What about some music, for instance?'

This was a far cry from the ecstasies of which he had been telling us; but Ranjit quickly sensed the quality and the potentials of the situation, and I could see he did not want to encourage Karam to make some rash offer which he would then insist on carrying out even if it caused much embarrassment to himself. He therefore immediately closed with my modest offer.

'Fine,' said he. 'Let's settle for that for the present. Let us have some good music, drinks, and some good food at Gyan's. That's a sporting offer!'

It was settled that the next evening, about seven, the three of us would meet at my house. I would provide the food and drinks, and Ranjit would produce musicians and a singer to entertain us. As he got up to leave, he kept saying, 'It will be a fine opening to the season, a fine opening!'

The next morning I was annoyed with myself. What had I contracted for? Ranjit was a very fine fellow, but I had acted absurdly to offer to provide the setting for the unknown propensities of his senses. Had I let him open a door through which he would lead me into places of his own choosing? I disliked the thought that it might be so. It would only make things worse now for me to cancel my offer; but I would take the occasion casually. I simply told my men that I might be entertaining a couple of friends, that there would be some music and we would dine when we felt inclined — perhaps there should be one or two additional meat dishes.

Karam was there punctually at seven o'clock, but it was not till seven-thirty that Ranjit arrived in his large roadster with an assortment of men piled into the back seat of the car. Jumping out, he said, 'Here we are, friends! I am sorry I am a bit late. We will give the musicians a few minutes to warm up while we begin to slake our thirst.' Ranjit had maintained his gay mood of the previous day, and immediately I felt more relaxed, telling myself that I was wrong not to have trusted his discretion in these matters. I forgot the rather tense half-hour of waiting with Karam who had not been in the most benevolent mood towards the absent Ranjit. 'Maybe he's gone off to Bhopal for the musicians, Gyan!' or 'He will probably turn up with a cartload of low whores,' he kept muttering sneeringly.

I was relieved when Ranjit arrived and brought with him, as arranged, just a few musicians. One of the gardeners, apparently aware of our festivities, came up the steps with several sweet-smelling jasmine garlands for all of us. 'Lovely, lovely! Flowers before all else! Scent is a promise, isn't it, my friends? A promise of more tangible things. A drink please, Gyan, my friend,' said Ranjit, boyishly enthusiastic.

We began to gulp the stuff down, when Ranjit looked around and called out to my men. 'A brazier, a skewer, and some meat à la brochette, please!' A man came out.

Ranjit told him to put several pieces of the spiced meat on the skewer and hold it over the open brazier. In a few minutes the sizzling meat was lightly barbecued and we ate the hot chunks, chasing them down with *thurra*.

We were in one of the two large drawing-rooms downstairs, sitting on the two chesterfields, several low tables in front of us, and the brazier, to the side, standing on a large slate slab. The musicians were entering the room: a *sarangi* player, stooping and old, carrying an ancient instrument with intricate ivory inlay on the shaft; a wiry young drummer with an alert smiling face on his long neck; and a third man, his eyes downcast and his heavy muscular body ominously alive. I wondered for what purpose he was there.

Last of all entered the singer. He was a small, very dark, smooth-skinned youth of about seventeen or eighteen, dressed in a black formal *achkan* and wide white trousers. His eyes were strikingly large and expressive, and as he entered there was a quick exchange of glances with the heavy major-domo accompanying the two musicians. The singer then turned to us and raised his slender well-formed hands in greeting. His mouth was full-lipped and shapely, and his eyes under their long brows lit up differently as his glance passed from one to another of the three of us on the chesterfields. The silent muscular man looked at him with heavy admiration.

The musicians nodded approval. The singer took off his *achkan* and displayed a liberally opened, white muslin shirt and an embroidered black and green silk waistcoat.

Ranjit now took over the situation. 'Rustam, young master,' he said, addressing the singer in a coaxing, almost affectionate tone. 'We are at your service. Come bind us with the spell of music, my friend.'

The young boy smiled and his eyes flashed. He cleared his throat and mopped his face with a colourful silk handkerchief. He whispered to the musicians and then holding one hand to his ear, he started softly intoning the air of his song.

Then came the first two lines of an old classical drinking song:

> 'What wine could it have been
> They poured in my silver cup'

Rustam sang in a beautiful baritone voice. He gave us the melody of the song, and then he trilled and improvised with delightful effect. 'Wah, wah ... wah, wah,' murmured the conductor approvingly. Ranjit repeated these cries of approval, and Rustam went on with his song:

> 'What wine, what wine, I ask
> That athirst for this fiery fountain'.

Karam was now gulping his *thurra* with zest while he devoured the delicious hot lamb brochettes. Ranjit was leaning back on the chesterfield, his eyes half closed and his face twitching gently with the changing cadences of the song.

Rustam sang on:

> 'What wine, what wine, I ask
> That athirst for this fiery fountain
> Furious god and angel join
> In immortal struggle with each other
> To possess this wine
> To possess this wine.'

He trilled fantastically, straining to fill the song with the high inebriation of the wine it celebrated. His drummer and violinist urged his skill on by finding enticing sidepaths of succulent improvisation. The notes entwined, sensuously piercing the very texture of life. I felt myself melt inside into a warm ether in which the music made its gyrations, runs, trills, and swift risings and dying falls. I glanced at my hands and legs to reassure myself that at least my exterior remained familiar and recognizable.

The song was over, and the musicians got up and silently slipped out for a smoke and some refreshment. The world tried to click back into its old focus, but two images now overlapped and physically hurt my blinking eyes. I looked at Karam and Ranjit. Karam seemed part of the old picture — his face was flushed with wine, and perhaps also by the music, but his eyes were cold and unmoved — while Ranjit had a soft passionate light in his eyes, his mouth was half open as if breathing the air of a different world and searching for a new language, and his ruddy face was now in glowing repose.

My men brought in dinner, setting large silver plates on the small tables set before each of us. Karam fell to immediately, smacking his lips and devouring the curried partridge and the saffron-scented rice. Ranjit, too, was eating heartily, but he sampled all the various meats, liking best of all the specially prepared dish of turnips and lamb. He was silent during the meal, but brightened up again towards the end.

'Music alone is permissible on a full stomach. Even love must wait, but we must have that divine Rustam sing to us of matchless tender love.'

'Yes, yes. Love I understand ... let him sing of love,' broke in Karam, lustily swinging himself full length on a chesterfield. Ranjit gave me a quick look, establishing between us our rather different mood and approach from that of Karam.

Rustam had now entered the room.

'Young master, are you ready again to receive us with music?'

'Sirs, you do me great honour. And I promise you I will be in better voice now. I have eaten only a mouthful or two, and slaked my thirst. My blood is flowing just right now.'

His musicians were seated, and the drummer was already slapping the taut skin of his instruments. Then he stopped to let Rustam intone the song. The boy's face was more passionate now. His lips were protruding a little and his eyes throbbed with flashes of varying intensity as the music welled up in him. Then he formed the words:

> 'I walk where every footstep is concealed
> I whisper where every sigh is muffled
> I do not know where I walk
> I tread the pathway of love!'

'Wah wah, wah wah,' called Ranjit. I, too, joined in. Karam was now sitting up, and, slapping his thigh with his large hand, kept time with the music.

Again Rustam sang, excelling himself, and the music blazed into the night. Confusedly I wondered whether the song was about me. I was powerless before the music. It encouraged me on into a strange world uninhabited by my past. Was there a glade of greenness, and flowers, and running water, before me? No. There was just nothingness into which I went, conscious only of a twin beat: the music's and my own — and of one word — love. But really there was nothingness. The music sped me on my inner journey. I wanted to go faster and faster in this land of nothingness for I knew I had to traverse the whole of it before I could gain the other side. And urgently I felt I had to get there, not knowing yet what it held for me.

I lost all awareness of Karam. Ranjit remained as a vague presence, a traveller in the same nothingness, but his path led off to one side — it seemed more verdant there, and the light shone through the trees above him, but it was all very hazy.

I was aware of my surroundings again. The large

muscular man was looking at Rustam with dark greed dripping from his eyes. I feared for the singer and instinctively turned to look at him. I saw the face of a singing angel — mysteriously feminine. Did I dare look at Karam? I fought against a dislike of the image in my mind of his ungainly figure and his unresponsive face. Now I turned towards him and thought I saw him also mellowing before the fountain of melodious sound. I was pleased that the evening was going so well.

Just then, Karam sprang from the chesterfield and sat down on the carpet beside Rustam, his huge body towering over the slight young singer. The burly conductor looked at him surlily, but decided that the moment of intervention had not come. Besides, Karam might become a valued patron and must therefore be allowed some leeway. Karam put his hand up to his ear and began trying to intone the air of the song, his eyes fixed admiringly on Rustam. But the singer was unperturbed. He went on singing, his eyes flashing at Ranjit and myself. Karam gave up his attempt at singing and began to stroke Rustam's back. The boy moved away. Karam followed him on his knees, his face flushed and his drunken arm stretched out unsteadily, reaching for Rustam's face, to chuck him under the chin. Rustam leaped up and fled. The burly conductor rose, and yanking Karam to his feet led him back to the chesterfield. He gave Karam a hard nailing look — 'Just you stay there', it said emphatically, and Karam looked suddenly deflated and ashamed.

The conductor brought back his singer and the song went on. At the end there was a pause. The conductor looked hard at Karam for a few minutes. My friend lay very still and appeared to have gone off into an inebriated slumber. That satisfied the burly man and he gave the musicians a sign to continue.

In the early hours of the morning Rustam stopped

singing. He was utterly exhausted by his effort and at the point of collapse. Till the end of the last note he had looked taut and glowing. Now, a second later, he was a shivering, cold, amorphous heap. His musicians put a fine Kashmir shawl over him; my man brought him hot spiced tea and meat, and in a half-hour or so, he began to come to life again. But now he was only a rather un-distinguished-looking young man — typical of the frater-nity of entertainers — swarthy, sensual, and underfed.

We roused Karam from his slumber. One of my grooms drove off with the troupe and Karam in a large phaeton. Karam had insisted on going with them, and the burly conductor seemed quite willing to have him go along. Ranjit was doubtful as to whether he should go, but decided that, at worst, Karam would make a secret assignation with the musicians to go by himself some evening to Rustam's dwelling.

Ranjit remained for another half-hour or so and sipped another drink. His face was now unbelievably different from what it had been when I first met him at Karam's house only the previous afternoon. Then his look of vigour, frankness, and geniality had interested me. Now all that strength in his face had been transmuted into a soft gentleness. There was a sad look in his eyes which glowed softly instead of sparkling as they had previously done.

Before he left he said to me, 'Gyan, you look sad, and your face has a soft glow which I understand well. I am glad we had this evening together.' He smiled, and as he briskly donned his coat he again looked the well-preserved sporting man hurrying home for a few hours of sleep before an early morning ride or shoot. That may well have been what he actually did, for the next evening he sent me two wild pheasant which he had shot.

*

More often than not, I too took an early morning ride now instead of engaging in a half-hour of sword play. The heavy morning air whirred past pleasantly, and the swift movement across the vast open stretches of country — dry to the west of Amritsar, and verdant and heavy to the east — with the sunshine just beginning to break through the morning mists, was often like an outer expression of my inner sense of a journey into the unknown.

Where was I going now on this inner journey? I did not yet know. For weeks and months I studied deeply in my library, and sought also to clarify my thoughts as I strolled on the lawn or in the garden. It was impossible to find my way. Where was I? Physically I was still part of the old family scene in Amritsar — the house stood over me as it had for most of the forty and more years of my life. But even here at home I had a bleak sense of remoteness. My elder son was now married and had his bungalow in the town. He looked after the paper factory and took trips to some of the big selling centres such as Delhi and Lahore. I saw little of him and his family. The younger boy was now in England. We corresponded occasionally, and I often wondered whether he, too, when he went to Oxford, would be mesmerized into a different world as I had been twenty years previously at London. His communications were dutiful but perfunctory, and I expected no more, for he had had little contact with me during the several preceding years as a student in Delhi. So I was alone in the big mansion.

Karam's relationship with me had been more successful when I was fixed in the ambience of my home. His passions, his warmth, his frankness, and his heavy blundering had amused me and, in a measure, endeared him to me; but to go to his home and be faced with all this in the raw, and to be in the centre as it were of his life and its consequences, was unattractive to me. When he had been a regular visitor at my home I had let myself imagine that he came because he sensed a richness and a rhythm which held him and amplified his own faculties. There was certainly no such spellbinding influence in his home to draw me. Nevertheless I did visit Karam and Sita from time to time, and we met more often in the company of his cousin Ranjit.

Ranjit gave himself fully in his relationship with me whenever we met. But it seemed necessary for him to be sporadic. We would see each other, have a drink, chat, and arrange to meet again soon to listen to music; watch the loveliness of some exquisite dancing girl; or perhaps drive to Jullundhur or Lahore and spend a day with my sister and her friends, or with friends of his at Lahore. But as often as not I would get a note from him saying he had had to leave Amritsar suddenly and would get in touch with me again on his return — and that would invariably turn out to be a matter of weeks, or even several months. It always interested me that it was not pressing business affairs that would suddenly take him away — though he did occasionally visit his lands near Ajnala — but invitations from friends all over the country. He had to travel. It was part of the steadying process that his intensely warm and emotional nature required. So, though our relationship was close, it was in only one small sector of the many-pointed far-flung pattern of life that Ranjit held taut under his restless feet.

Strangely enough, and it was strange because it sprang from so fortuitous and unequal a beginning, it was my relationship with Narinjan Das and his family which most successfully carried over from the previous period of my married life to the next phase. Circumstances had deepened the relationship. After allowing me six months of solitariness following my wife's death, Narinjan Das again took up his old habit of visiting me almost each morning. Since he was now in full charge of the timber yards as my partner in the venture, and as I had given up taking an active part in the business, it was no longer incumbent on him to visit me. But still he came, and, without inflicting the burdensome minutiae on me, he kept me in touch with our affairs. It was, however, primarily his own life that he brought nearer to me.

Narinjan Das was full of patent contradictions without which he probably would not have found it possible to sustain life. I recollected how, almost twenty years ago when I first met him, he was so assiduous in his attentions that he would always come running when I called, jumping over logs of timber or stumbling across potholes or open drains which he refused to avoid because that would have meant deviating from the straight course. Often he hurt an arm or a leg in this display of intrepid devotion to me and to his duty, which was in sharp contradistinction to his personal timidity in normal circumstances. For instance, he never came within ten yards of my horse lest the animal should shy at him. Smiling, he would say, 'Sir, life gives one enough kicks without offering oneself to the hoof of your noble beast!'

With all his astrology, contradictions, and apparent obsequiousness, Narinjan Das was a highly intelligent, amusing, efficient, and staunchly loyal person. I had found him increasingly likeable.

About a year and a half after my wife's death he came to me with a shy, flushed look. His eyes were bright but coy. I wondered what could have happened.

'I am glad to see you looking so happy today. Are the stars favourable to you, Narinjan?' I asked teasingly.

He gurgled with soft laughter, an expression which he seldom permitted himself in my presence. 'Yes, sir, very good stars are rising. A very happy event for me is about to take place. Great happiness is coming to me.'

He must have been in his early fifties, about ten or twelve years my senior. I was very well preserved. My body was firm and strong, and the thick hair on my head had not begun to be streaked with grey. But Narinjan Das I somehow regarded as a little decrepit: he was pot-bellied and flabby. I never thought of him as a dashing romantic figure. But what was this he was saying? It could only mean that he had decided to marry again. I smiled warmly, and putting my hand on his shoulder I started to say, 'Congratulations, Narinjan. I am sure it will make your old age very happy to ...'

He broke in, not able to contain his news any longer. 'Sir, I knew you would at once follow. You are so wise, you always know. My daughter, Indira, has been accepted in marriage by Dr Santokh Singh, a very prosperous young doctor in Jullundhur. It is all due to your goodness and kindness to me that this has happened!'

How little I *had* known! Narinjan continued telling me of his happiness, and asked me to be father to him and accept responsibility for the arrangements for the marriage. I readily agreed, and it was decided that the festivities would take place at my home which would temporarily become also the Narinjan Dases' home. Since it was only about a month to the marriage, he often brought his daughters to the house in the intervening period. Indira was a handsome girl of about twenty-two. She was extremely poised and well mannered. Narinjan Das had given her an excellent education, and she had just completed her Master's degree in Delhi. I had not met Santokh Singh, but I felt certain that it was he who had done well in this match.

But it was Indira's younger sisters who were strikingly beautiful. Pushpa, the second girl, had a breathtaking quality. Her eyes were unbelievably large and clear, and her olive skin seemed to vie with the clarity of her glance.

Her face was oval, her mouth perfectly formed, and her nose in perfect proportion. Her hair was glossy and thick. She was of medium height, and her body was superbly made. Each of her movements was naturally as graceful as if she had been the greatest dancer at the Mogul court. And to all these superb physical endowments had been added a beautiful soft voice and a gentleness of response.

The youngest girl, Sulochna, was still only fourteen or so, but she bid fair to be able to stand next to her sister Pushpa without fear of comparison. She had perhaps a more fiery look in her dark eyes under slightly heavier brows than Pushpa's. Otherwise she was remarkably like her sister.

It was a pleasure to have these beautiful girls about the house. The rooms came to life again. Indeed I had all the carpets and covers cleaned, and some beautiful pieces which had been lying in storage were brought out to add their vividness and warmth to the splendour brought to the house by its new occupants.

It was in those days that it occurred to me that as Ranjit and Narinjan — and now his family — were my closest friends, I should attempt to bring together the two very different planes of life on which I lived with them. At first the idea seemed to be too daring. How could I bring the timid orthodox Narinjan and his lovely daughters together with Ranjit, who was primarily, in spite of his great sweetness and warmth of character, a highly individualistic hedonist? What about our mutual epicurean pursuits in the realm of song, dance, good music, good food and drink? At first sight it was preposterous that I should think of bringing together my two very different worlds.

In this state of mind we are all quite capable of creating accidents which put our confused wishes to the test without seeming to involve us in any direct responsibility. On one of the evenings before Indira's marriage, Ranjit and I were to get together to hear Rustam render some new songs that he had been learning. Caution would have demanded that we should meet at Karam's or at Ranjit's own town house. But when Ranjit said we would meet at my house I raised no objection.

The evening arrived. Narinjan noticed, as must have his daughters, that special preparations were being made for guests at dinner. Normally they ate with me before returning to their own house for the night. On this occasion Narinjan said, 'Sir, as you are having other people over tonight, please permit us to leave early.'

I waited a moment and then said very quickly, 'No, no, my dear Narinjan. You must be here as usual. You see, I have a friend who enjoys listening to good music and we are having a singer over tonight to entertain us.'

Narinjan looked at me, his eyes popping and blinking rapidly.

'But, sir, in that case how can I, with my daughters, be here?' he said in a timid and tremulous voice, while his hands fidgeted nervously.

'Nonsense, my dear man!' I said firmly. 'It is perfectly all right if they remain. The singer is a young genius called Rustam, and they will enjoy his music. Besides, Narinjan, I am here to look after them.'

Narinjan drew himself up when I uttered the last remark, and his face, till now contorted by timidity, became resigned. Quietly he said, 'Sir, you are my protector and my father. Of course we are safe with you. We owe everything to you. And as it is a man who comes to sing, perhaps the girls can stay for a short while.'

Had I been wise? I wondered. And as if to cut short my doubts I added with a quick smile, 'Yes, just for a short time let them stay, Narinjan. Then you can take them home. It will be best that way.' I felt I had hit upon a reasonable solution, and I remember congratulating myself that I had done so well in the difficult situation which had been thrust upon me.

When Ranjit came that evening, and we were waiting for the musicians to arrive, I told him that I had a surprise for him.

'You know my man, Narinjan Das. You have often heard me speak of him?'

'Yes, of course. The queer astrological character. Don't tell me he's here!' was Ranjit's reply.

'Yes, he is. He is not so bad. You know he's quite a good friend of mine and I thought you two should meet.' I was deliberating while I said this as to whether I should also mention to Ranjit about the presence of the daughters. But he cut me short by saying, 'My dear Gyan, if you think I should meet him, I am happy to do so. I hope he is not the kind who will put off the singer. You know, there are types before whom a sensitive artist will not perform.'

Poor Narinjan probably was that type, but I had in mind his daughters when I said, 'N — no, Ranjit. No, I don't think Rustam will be put off. Perhaps to the contrary!'

We poured ourselves some *thurra*, and when Rustam and his accompanists arrived Ranjit was telling me that any day now he was about to leave on an extended trip to North Bengal tiger-shooting and spending some time in the hills. He expected to be gone for at least two months.

Now that events were leading up to a meeting of my two worlds, I felt acutely put out by this announcement. I wanted Ranjit to remain at Amritsar, for some time at any rate, to help me bring the two worlds together.

I said to him, 'What a pity. I would have liked you to be here next month when Narinjan Das's eldest daughter is to be married at my home — he so much wishes to have the marriage here and I do not want to disappoint him.'

Just then, Narinjan Das, who must have asked my men to inform him of the arrival of the musicians, entered the room with his daughters. I noticed that Ranjit's glance had picked out Pushpa, and that his face spontaneously expressed pleasurable surprise and more than a touch of incredulity. Meanwhile, Narinjan Das had acquired an over humble expression, in keeping with which he quickly raised and folded his hands to us in greeting, keeping his head bowed while he did so. His blue serge suit was clean, but somehow in the context of his behaviour it looked very subordinately functional. No one could have imagined that the three girls next to him were his daughters. I had not planned any formal introductions, but wanted to say something to encourage Narinjan Das to alter his bearing.

'Come, Ranjit,' I said. 'Meet my *partner*, Narinjan Das, and his daughters — Indira, Pushpa, and Sulochna.'

Ranjit came up to them with me and greeted them courteously, and either because he liked the look of my guest, or was reassured by my introduction, Narinjan Das's face beamed and he lost his obsequious look and manner.

My men had some refreshments ready for Narinjan Das's family. We sat down and the music commenced. Deeply moving though Rustam's execution was, I could not help observing my guests. Narinjan Das sat stiffly in his chair. His objective was to look respectful, and he was achieving it; but on observing him for a few minutes I realized that under the over polite mask there was a steady preoccupation with other things. At one point I had no doubt he was figuring out when he could gather his children together and leave without giving offence to me and to my other guest.

Among the girls, Indira's face alone remained inscrutable — dignified and poised, as though the sounds as they permeated all fitted into a prearranged pattern in her mind. As I looked at her it struck me as being an inexhaustibly capacious pattern, and, on that account, admirable. Sulochna, the young child of fourteen, seemed wonderstruck at the virtuosity of Rustam. Her eyes were very bright and concentrated, and a flush had risen to her cheeks. She sat with her hands clutching tightly at the arms of her chair, and at times aided by the angle of the light, I could see that her skin goosepimpled whenever the music was very intricate and Rustam's voice pirouetted and improvised around it with power and fluency.

It was Pushpa, though, whose face was the answering masterpiece to the superb singing of the young master. She sat straight in her chair, but the even glow on her face, and the easy gracefulness of her arms and hands which occasionally she moved from her lap to the arms of her chair, indicated that she was not doing so out of tenseness. It was part of her responsiveness to the music. Her lips, slightly apart and fuller than normally, seemed to be drinking in the sensuous stream of sound; and the music seemed to induce the light to play with her face as if soft downy fingers were touching it. Her eyes shone with deep, or soft, or scintillating, lights according to her varying response to the music.

Ranjit, in a characteristic pose, leaned back against the sofa on which he sat. When he listened to music, generally his eyes would be closed and his face could be seen twitching gently in response to the singer. Today his face was very still, and his eyes as he leaned back were opened — just a slit, but enough for me to see that they were fixed on the beautiful Pushpa. Why not? was my immediate reaction. The girl was so unusually beautiful that one could not but look at her. But as my own eyes wandered round the company, including the singer and his musicians, I realized that Ranjit was not merely casting an occasional glance at Pushpa. His eyes never left her, and the expression on his face, unresponsive to the rise and fall of the music, was of a deepening intensity which showed in the rising flush on his cheeks.

After a half-hour or so I noticed that Pushpa had become aware of Ranjit's gaze. Apparently it disturbed her. Instinctively her eyes fluttered in an effort to brush aside the disturbance and to persuade Ranjit's eyes to abandon their stare. But Ranjit either did not wish to or seemed unable to stop gazing at her. After another few minutes the girl apparently decided to seek assistance. Shyly she began looking at me, appealingly, for a moment at a time. I kept trying to catch his eye, but Ranjit's gaze was unalterably engaged and my glances passed unregarded.

Rustam finished his song and I thought I could now get up and break Ranjit's concentration. But spontaneously we all cried, 'Wah, wah, wah wah', and this encouraged Rustam to repeat his favourite portions of the song. I was absorbed again in the music and forgot that I was being asked to accept the responsibility of diverting Ranjit's attention from Pushpa.

When the encore was over Rustam was exhausted, and his conductor handed him a large silk handerchief to soothe his excited face. I got up to order some drinks, and it was not till I turned round that I found that Ranjit had gone over to speak to Narinjan Das. He was saying, 'Your daughters seemed deeply interested in the music. Perhaps they would like more opportunities to listen to great singers and musicians. That can quite easily be arranged, and I hope they will come.' Ranjit, who was well schooled in the conventions, spoke very courteously and his face was gravely serious — as if he were talking about an important duty which had to be fulfilled.

To my surprise, Narinjan Das raised his face with a sort of unflinching courage against Ranjit's statement, as though the latter had been belabouring him with blows rather than inviting his daughters to a pleasant evening of entertainment. He said, 'It is very kind of you, sir. My daughters are always attentive, but they are very busy and may not be able to come, but it will be as Rai Sahib, their patron and my protector, wishes.' And he turned to look at me for my adjudication.

It was as if Narinjan realized that though Ranjit was the type of man who behaved according to the conventions, he would not limit his life to their narrow restrictions. Again I was being appealed to on behalf of the girls.

I smiled and said rather abstractedly, 'Of course. Music is a wonderful thing. Whenever they can spare the time they should certainly hear more of it. Come now, let us eat some dinner, my friends.'

We sat at our low tables. Pushpa came and sat by my side, and I saw the disappointment on Ranjit's face. But that was only momentary. He sat beside Indira and led the conversation all through the meal. He asked after her studies, then he told of his own interests, his travels, and now his curiosity about the young science of anthropology. That was what was taking him to the hills of North Bengal and Assam. From time to time he glanced at Pushpa to see how she was reacting to the exposé on himself. It was obvious from his glances — and undoubtedly she realized this — that his statements were mainly for her ears. But the girl gave him no encouragement. She conversed with me about her sister's coming marriage, and how happy it would make her father. Because he had had no sons himself he was hoping that his daughter would present him soon with a grandson. Then he would feel that his duties as progenitor of the family had been performed. All this Pushpa talked of with understanding and detachment.

All through the meal she gave no indication of her reaction to what Ranjit had been saying.

After dinner, Narinjan and his daughters left. Ranjit filled his glass with *thurra*, and called upon Rustam to sing a song 'dripping with sorrow'. He gulped his drink and said, 'I want to hear of a deeper sorrow than has ever been told, something that will overwhelm my own sadness. Come, Rustam, help me!'

The young boy smiled pensively as if to indicate that he, too, knew the sadness of love. Then he intoned his song in a melancholy air that came, he later said, from the sad bare hills of Rajasthan.

At the end of the song Ranjit managed to look rich with misery. I tried to cheer him up. There was, after all, no reason for his deep gloom. Indeed, I was surprised that this highly sophisticated man of well over forty should be affected so instantaneously and powerfully by a girl of twenty. And least of all had I expected Ranjit to show his emotions so blatantly. When the singing was over, Ranjit — still looking dejected — got up and left abruptly, as though he had realized suddenly that he was late for another appointment.

The day of the marriage came. Preparations were being made to receive the bridegroom's party in the early part of the evening, and for dinner and the marriage ceremony which were to follow. To my surprise Ranjit arrived during the afternoon. I had not seen him since the last musical evening and had assumed that he was away in Bengal. His eyes were restless and his lips bloodless — as though he had been living through a period of strain.

I was genuinely happy to see him again. 'My dear Ranjit! What a pleasant surprise. I had feared you would be away in Bengal. What happened?' I asked with some anxiety as he came right up to me and I could see how ravaged his face had become. 'Are you all right?' I added.

Ranjit smiled rather wanly, but his voice was resonant and firm as he replied, 'I am all right, Gyan, and I hope I am not intruding today. I had to postpone my trip — I did not feel like all that nomadic wandering in the hills.' I assured him that everyone would be delighted to see him.

We were awaiting the bridegroom's party. Narinjan Das stood uneasily, looking as if we were approaching the edge of a precipice, and glancing at me for protection. Ranjit and I both tried to suppress a smile.

'Do calm him down, Ranjit; I will have to see to the other guests,' I said to him. 'Narinjan Das has put everything on my shoulders. He insists that I am his father — which means that dear Indira is my granddaughter!'

Ranjit responded to my suggestion immediately. At first Narinjan looked even more uneasy when Ranjit tried talking to him, but soon he relaxed and they were laughing together — a rare mood for Narinjan. The whole evening Ranjit was assiduous in his attention to Narinjan, which also gave him a vantage point with the younger daughters, Pushpa and Sulochna, who kept coming to talk to their father. Later, Narinjan said to me in his characteristic fashion, 'Rai Sahib, Sardar Ranjit Singh is a great gentleman, worthy of your noble friendship. I am greatly indebted to him for all his help this evening.'

After the ceremony was over and the guests had departed, Ranjit was getting into his car. He looked very weary and said quietly, almost submissively, 'Well, Gyan, I have tamed Narinjan. But what an effort! And the girls, I think, would like another musical evening. Let us arrange something for them.' He added this last point in a tone which suggested that he had discussed the matter with Narinjan and was passing on the result of his own considered reflection.

It was clear that Ranjit wished to get to know Narinjan

and his daughters. It was no surprise to me when, three days after the wedding, he sent word that he had arranged for Rustam and another singer to come to my place on the following Tuesday. Narinjan and the girls were still spending much of their time at my home and I told them at once of the arrangement. Narinjan looked wary, but the girls were obviously pleased.

In the next few days, Pushpa — less busy now — often came up to me as I strolled in the garden, and once she visited me in my ground floor office. She spoke of her plans. Now that Indira was married she hoped her father would let her study the things she liked — music, drama, and Indian classical literature — and also permit her to travel and meet people. Being at my house had been a tremendous experience: she loved the beautiful things, the books, and many of the people she had met. It had fired her with a desire for much more than she had previously ever dreamed of — travel, learning languages, and she hoped some day she might write a book. She addressed me for a moment with an appeal in her eyes, and then hastily turned away — 'But I don't want to leave Amritsar; I don't want to leave here.'

Before I could say anything she turned to me again, her face radiant now and her voice breathless, as if she had made a discovery. 'I know! Please let me come here and read in your library.'

I, too, felt I had discovered something, and pressing her hand I nodded, saying 'Yes yes, of course, my dear.' Were we both clinging to the same thing?

Narinjan now had a car and chauffeur, and it was easy enough for Pushpa to run down at least once a day to visit the library. That became the settled routine; and each succeeding day she would spend more time talking to me. Occasionally she was prevented from coming, and I would be restless and unable to turn my mind to anything. On the day after such an omission, she would linger longer with me in the garden or in my office, and she would look at me with a special warmth as if to make up for her absence. I tried not to admit it to myself — the girl was less than half my age — but it *did* seem that we were clinging to the same thing.

Our most certain expectations can be belied. I never doubted for a moment that the evening of music planned by Ranjit would materialize — he was obviously keen that it should. But the day before we were to meet he sent word that friends had come to town and he would have to call it off. For two weeks I saw and heard nothing of him.

Pushpa did not appear to be disappointed. She hinted that it did not matter much now. She had other dreams. She wanted to go to the ends of the earth, to learn all she could, see all the paintings and sculpture in the world, and meet all the interesting people; and then she wished to come to rest — right here in Amritsar — with all her acquired gladness and pain of experience. She was lyrical about the town — the golden temple of the Sikhs, the old houses with their timber balconies, the crowds in the streets — restless and rough and yet always courteous to women — and most of all, she loved the wide plains around Amritsar. She had been going out in the car and driving for miles — 'so much space in which to let one's heart go, and then to feel it return all filled with life,' she said — and the distant romantic fringe of mountains called her to its own deep intricacies. 'I could lose myself for ever in those mountains.'

Not that she consciously sought a balance — but she had it in perfection between her sense of adventure and her absorption in the life of the town. She was always responsive and eager when I talked of my own travels. Why did I not travel again, she asked. I shook my head in reply. I did not wish to say. It was too complicated to explain. I had little inclination now to travel. When I was alone I often asked myself why not. Did I cling to the old house? After some time, I thought I knew the answer. I had to besiege the house, for still it remained unconquered. Here I had lived most of my life, but the place had remained inviolate. The secret room was a symbol of what I had yet to realize. It had never come alive for me. Except at the very end, my wife Basanti had never liked the room and it had remained barren and unreal. In a sense the whole house had remained that way.

A few weeks later Ranjit sent a message to say that he was now free again and we could meet the following week. This time he kept his word. There was another singer — Rustam's teacher; an old master who could simulate and outdo the fires of youth. Each man sang more passionately and perfectly in response to the other.

Again Pushpa sat near me, and when the music rose to great heights of intensity she would look at me, and I noticed, too, that her hands were restless. Ranjit fixed his powerful gaze on her from time to time, but he seemed more resigned than he had been on the previous occasion.

But at dinner he pulled up a chair near us and talked to Pushpa and myself. The music, he said, had roused him from his lethargy, and he would hasten his preparations and leave on the trip he had planned to the north-eastern part of the country. He might even go through Sikkim into Tibet — to Gantok and to Lhasa, if he could get permission from the Dalai Lama. Some years before he had paid a fleeting visit to Sikkim and the calm-faced lamas in their long robes of deep glowing red had fascinated him. They had shown him the *thankas* — their silk banners painted in the quiet winters in the monasteries. Intricately, and in every variety of colour, they had illumined

127

on the silk the sacred stories of the life of Buddha. Though the subjects were set, and the patterns of the paintings conventional, he told us how he was able to detect the greater artists from the lesser. From a certain hand the colours had glowed and were three-dimensional. Into the details had been incorporated little touches which the artist could later explain as slips, but which in fact his own vision had demanded of his brush. Perhaps as a penance he had had to turn the prayer wheel an extra ten thousand times, but while obediently he performed his penance he was already seeing the vision of his next heretical but beautiful *thanka*.

Pushpa was deeply stirred by Ranjit's talk. He was aware that she was under his spell and wanted to see how far he could sway her.

'Would it not be wonderful if we all could go?' he exclaimed.

Spontaneously Pushpa responded, her eyes flashing and her hands clasped in her excitement. 'Yes! Let us go! Please, let us all go!' And she gazed at him for a long moment pleadingly.

Narinjan Das shifted uneasily, and that probably kept Sulochna from endorsing her sister's wish. Ranjit smiled to himself and decided to press the point further. Without making it clear as to whom he was addressing, but looking first at Pushpa, he said, 'Then come by all means. You are very welcome.'

Pushpa opened her mouth, but for some reason she glanced first at me, then at her father, and decided to say nothing.

As the old master could no longer keep late hours, there was to be only a short spell of singing after dinner. We prevailed on Narinjan to wait till it was over.

The best had been reserved till the end. Beautifully the old master sang a love poem by Ghalib. It was resonant with sadness as only Ghalib could make sorrow sound: a tempest of tears that was also balm to all the open sore places of life.

When Narinjan and his daughters got up to leave, Ranjit said laughingly, 'Well, unless you see me in Tibet, it will be a long time before we meet again!'

I could see that Pushpa was very agitated by his remark. It was raining when they left. Ranjit mustered a couple of umbrellas and accompanied the girls to the front veranda. I went out with them and saw how closely he observed Pushpa's agitation as he helped her into the car.

That whole evening Ranjit had been debonair. But as soon as the others left his earlier mood disintegrated and he was overcome by melancholy. I gave him a drink but it had no effect. He sat silently, his eyes half closed, and his body slouching in an armchair. Some great weight seemed to press upon him.

About half an hour later he revived. He stood up and shook himself. Then turning to me he said, his words coming rapidly:

'Gyan, come with me tonight. It is intolerable for me to continue this way any longer. Don't ask me yet. Just come with me.' He stood before me, appealing to my sense of friendship. He looked helpless and in great need. I did not inquire where we were to go. As we put on our coats I asked, 'Shall we take my car?'

'No, no. I will drive you,' he replied, very definitely as if to make it clear that my car would be inappropriate.

We got in his car and drove off. I had raised no questions, but I had the uncomfortable feeling that I was being called upon to interfere in someone's life, and I thoroughly disliked the idea. I struggled against my suspicion and managed not to blurt out a protest. We had driven into the old city when Ranjit asked quietly, 'You have heard of the Maid of Amritsar?'

I was not sure that I had, but said, 'Ye—es', vaguely.

I had thrown Ranjit's calculations out, and he hesitated. But then he went on, 'Well, you will see her now. She is reputed to be the greatest beauty since the fabled Ranja.'

We drew up in a narrow street, before a small wooden door in an old brick wall. Ranjit got out and knocked. A man whispered from within, and on hearing Ranjit's voice he opened the door. We entered a pleasant little courtyard, a lamp lighting up its miniature fountain in the centre, two flanking cypresses, some shrubs and beds of flowers. Through a veranda we entered a sitting-room furnished with low mattresses and long bolsters all up-holstered in embroidered pale blue silk. On the walls were several photographs of a beautiful woman, framed in varnished mahogany-type wood, and similarly dis-played were many scrolls of adulation addressed to 'Askari, the Maid of Amritsar', emblazoned in Persian, Urdu, Hindi, Punjabi, and English.

Hearing the tinkling sound of a woman's jewellery, I turned away from the scrolls on the wall and saw a woman entering the room. Immediately she greeted Ranjit with joined hands. He went up and put his arm around her — rather perfunctorily — while he asked tenderly after her health. Then he brought her to me.

'Meet my friend, Gyan. You will like him,' he said.

She again raised her joined hands to greet me. Then she smiled and looked up at me.

Under her long black eyebrows, her almond-shaped Persian eyes had been touched by the darker hues of the Indian plains. They were translucent grey within a fringe of black. She had all the other attributes of striking beauty: glossy hair, shapely ears, a fine yet sensual nose and mouth, a not too narrow lower jaw, graceful hands, a buxom figure, and the much sought after light golden complexion. Her glance had a complexity which I had never seen before — with a disconcerting sauciness it combined innocence, but there was also intelligence and fire.

I looked for a moment at Ranjit. He was proud of her, that was clear; but also tired of her.

Two women brought us some refreshments. While she was serving me, Askari said, 'Of course I have heard of you, and I am glad you have come at last. I only know his friends as those who take him away from me.' She did not even look at Ranjit as she thus taunted him.

'Well, I can't be in two places at the same time. And when I am not with you it is surely better to be with friends than with enemies,' he said gently and persuasively.

She smiled and looked at him now. 'You always talk so beautifully. I forgive you; and more so because at last one of your friends has come with you. But tell me. This does not mean that you love me less? — only that you want your good friends to know me?'

Ranjit sighed. His eyes looked very calm and calculating — as though he was raising his sights to a distant target. 'Of course I want my friends with whom I spend my time to know you, and you to know them. Then you will be able to pass the time happily when I am not at Amritsar. You know I have to leave you frequently to go away and work!'

'To leave me, and to work! Could you not work in Amritsar? Look at your friend, Gyan. He does not seem to me the sort of man who leaves Amritsar to work.'

I sat up straight. How did she know that? Had she just chanced it, or did she possess a special instinct which laid a man bare before her?

'Yes, yes. Men do all kinds of things — some this and some the other. Ask Gyan if my work and studies do not take me far away. Ask him if I have not to go to Bengal and Assam, and perhaps even farther, to Tibet?'

I felt rather uncomfortable, for I had not expected to be cross-questioned on this matter. But Askari was thinking of her own life. 'While I will go from room to room to watch your distant travels. One room will fill me with fear — like the tigers of Bengal; another with loneliness, like the wilds of Assam; and the third with the dizziness that will overtake you on the jagged peaks of Tibet. So I, too, will travel here and be desperately exhausted when you return to Amritsar! What fate is this for she whom men call the sister of Ranja, the Maid of Amritsar, Hafiz's transplanted rose of Ispahan!' She said this dramatically, but with real feeling, and I did not know what I would say if she were to ask me whether Ranjit really had to go to Bengal and to Tibet. But she was in no mood to question me. After her speech she sighed and sat with her head buried in her hands.

A nervous silence followed. I glanced at Ranjit. He was suppressing a frown of boredom, and in his look was the subdued brightness of the confident gambler. I tried to forget this strange drama as I nibbled slowly at some fruit.

After an interminable time I heard the tinkle of Askari's bracelets. She was on her feet. I looked at her. She was smiling now, and her face was no longer self-absorbed.

'We must not rattle our noisy sorrows in the presence of your good friend. I can see he is a gentleman of sensitivity.' She served me again with fruit, and then addressed a question to me. 'Don't you think, sir, the air of Amritsar is sweet? Do they not say that only the very wise live in this town, and because there can never be a large number of very wise people this will never be a great city like Calcutta? I can see that you belong to the select few.' It was a pretty idea, and I could not help smiling. I looked at her. She wanted some affirmation from me, a response to something deeper than what she had said. I was puzzled as I replied to her words.

'I do not know whether the air is sweet, but I do like Amritsar. Unlike my friend, Ranjit, I do not crave travel — and work that takes him to distant places.' I added the last few words to conform to Ranjit's interpretation of his absences, but I did not altogether like myself for doing so.

Askari looked at Ranjit. There was a remote expression on his face as he sat with his head tilted back a little, gazing across space as if he were surveying some distant scene. I then realized the deep restlessness of Ranjit's nature, and recollected that at my house, too, he had often sat leaning back, his eyes half closed and his face brushed lightly by the twitching air of some distant region. Seeing his abstraction Askari said slowly, each word carefully enunciated but not with any bitterness against Ranjit:

'Then, Gyanji, since we both live in this city of the wise, perhaps our paths will cross sometimes.' She looked at me appraisingly, as if to reconsider her deduction, and then she nodded and smiled, pushing up her lower lip playfully. She was pleased with the idea.

Ranjit still looked remote and he seemed not to have taken in the import of her remark. But the meaning had penetrated all right, for he seized this moment to make his really crucial announcement to her.

'So, Askari, in the next few days I must leave Amritsar. I do not know how long a time my travels will take; and since they are undertaken in connection with my work, and for understanding more about life, you will not expect me to try to circumscribe them precisely even before I have set out on them. In the past I have been able to tell you when I would return, but this time do not expect that of me. As always, you will be well looked after.' He looked at her with a kindly but somewhat troubled glance, not knowing how she would react.

133

She sat silently for a few minutes. Then she raised her face, and not looking at either of us she said, 'You go away for knowledge. I understand that. But, here, what knowledge do you apply? You tell me that as always I shall be well looked after. Do you not know that when you go I never am, and can never be, well looked after? You turn your back on these matters, which concern me, to dabble in others. Maybe the air of Amritsar only conduces to understanding in some. I do not know. But, you see, though we quarrelled the last time you were here and you hinted that you would soon leave me again, now I cannot even argue with you. I listen to your plan and that is all.'

Ranjit shifted uneasily, but his brow was untroubled again. Clearly he had wanted this meeting to be a peaceful one, and he was relieved that she had taken his announcement without reacting violently. He had dissembled both with her and with me by the pretence that his work was taking him away. He had expected her to see through his story and to hit out angrily, but she had maintained her poise, and as a result his own performance showed up very poorly.

I felt badly for him. He had brought me with him to lean on me, and to join my assertions to his so as to convince this beautiful woman that there were compelling reasons for his leaving Amritsar. Knowing I would have hated the false words I uttered, he must also have realized that to make this demand of me would have meant a big gamble with our friendship. This knowledge of his desperate plight somehow filled me with compassion for him, which was not diminished by my respect for the poise with which Askari faced her own torments.

Ranjit rose to his feet. In a flash Askari came over to me, and, smiling, she said, 'Must you go too, wise man of Amritsar? I am in the mood to talk of the complex knots of life, and I feel you would help me loosen them.'

I too had risen, and I bowed to her. It was not because of what she said, though her words provided the occasion, but because there was in me a pent up, spontaneous, and overdue acknowledgment of her beauty, her restraint, and her understanding. I excused myself, glancing at Ranjit, and Askari did not press her rhetorical invitation.

We left, and Ranjit heaved a sigh of relief as we got into the car. I too sighed, but for a different reason. He turned to me, and, nodding, he said, 'Well, that is over for the present. Thank you for coming.' Then he added, the thankful soft look in his eyes giving place to a victorious brilliance, 'Beautiful women have the right to assert their superiority. I never argue with them. When I fall in love I tell myself the fall is complete — not just emotional but physical and mental. It's humiliating in a way. You must forgive me, Gyan, for having led you into this today, but I felt I could not face her alone.'

It was a brave attempt, now that it was over, to proclaim that it had all gone off in accordance with his philosophy of life. It did not impress me, but nevertheless his new buoyancy removed some of the weight of uneasiness I felt. As this happened I felt my mind filling with thoughts of Askari. Who was she? Frankly, I had never heard of the Maid of Amritsar. Was she a woman Ranjit kept? Or was there even some closer understanding or contract between them? And why was he so obviously bored with someone who struck me as remarkably interesting, in addition to being superbly beautiful?

In his elation Ranjit drove very fast, and almost immediately we were at my house. It must have been three or four in the morning, but my head was so filled with questions that I wanted Ranjit to stay, and I asked him to come in for a drink.

'Yes, I feel like one!' he replied eagerly.

When he threw off his coat and took his drink, Ranjit

was no longer the distracted keyed-up person who — a few hours ago — had taken me from this very room to see Askari. His eyes gleamed, there was a triumphant smile on his face and a jaunty freedom in his movements. It was the mood in which a man sees fit to praise himself. He was bursting with it, and I let him talk. It turned out well, for much that he said bore on the questions that I was repressing.

'Four years ago there was something like a *swayambra* here. Askari was already famous as a beauty, and though she refused to subject herself to the full and strenuous course of training prescribed for one aiming at becoming a great virtuoso, many people considered her singing exceptional. When her mother, a peasant woman from the north, let it be known that she would welcome young men interested in Askari, there was no mean stream of callers. It was like the opening of the share market in Bombay for a new stock, if you like; or, as I said, it was a *swayambra*. Their house was besieged by throngs of men — burghers, prosperous neighbouring farmers, shopkeepers, some up-country landlords, Marwaris from Jaipur, and a number of rich princelings, even the heirs of two maharajas, and many young men with enough money to indulge their fancies. The old woman was surprised. She had not expected this torrent of interest. Had only a few honest solid burghers collected, her task would have been simple. She and the headstrong young Askari would have chosen the one that appeared likely to make the best husband, and could produce fitting proof on the matter — such as the attested extent of his worldly estate.

'But the situation was too complex for a simple decision, and Askari's mother seemed to follow a train of thought something like this. She herself had been a peasant girl, beautiful and bold, who had ensnared in an affair one of the richest landlords of the Punjab. As a result, Askari was born. She had demanded her rights. Before the civil courts, she would have had a very strong case; but, as she was of lower caste, under the customary law of the locality she was in a bad spot. The landowner offered her a handsome cash settlement, and bought her a house in far away Amritsar — that was how the mother and daughter came to live here. Askari's mother always played the part of a great lady. She felt she had done very well for herself and saw no reason why she should not continue to do so.

'Seeing the great throng of suitors who came to the house, she decided that if Askari would only agree to behave somewhat unconventionally she might get much more out of life than a humdrum existence with a solid citizen of the town. She persuaded the girl that it would be much more rewarding to enter into an arrangement with a man of wealth and culture, rather than marry a common fathead. Askari readily agreed — she was high-spirited, had great confidence in her extraordinary charms, and had no love for a life of bourgeois dullness.

'The mother screened the aspirants. All the ordinary ones were turned away. About a score of princelings, landlords, and merchant princes remained. Askari took one look at the grandees of the cities and all of them had to be struck off the list — which left the situation in the hands of a few great landlords and younger sons of our northern princes. The old woman was wise. What was a younger son of even a king compared with a landlord in his own right? Besides, it was a landlord who had escaped her, and it was among them, therefore, that she was determined to find her quarry — so the junior princes were shown the door.

'By this time, the half a dozen of us that were left' — this was Ranjit's first indication that he was among the contenders — 'were in a state of frantic elation and expectancy. We were the few who, in a series of screenings, had been

137

selected to survive in this great contest. As our excitement reached its climax, Askari seemed to wax in beauty and poise, and each of us thought himself a Hir or an Apollo. We were young men who had started out by fancying ourselves very highly, and at this advanced stage of the contest we regarded our special worth as having been established incontestably. Some of us were wealthier than the others, but we were all men of considerable substance and of respected families in the neighbourhood. None of us was town bred. All our estates were in the country, and in a sense that made it easier for us to subject ourselves at Amritsar to the situation in which we had placed ourselves: on our own rural estates it would have been impossible. This point, I suspect, the old lady did not fully grasp.

'For ten days we were kept on tenterhooks. During those days, two — I think the third and the seventh — I spent with Askari and her mother. My memory may be playing tricks with me, but as I remember it I did not put myself out very much. I was casual but — and you will forgive me for saying so myself — I was charming, handsome, and rich. After the ten-day period, the mother sent for me and said that without hesitation her daughter had selected me. The old woman shrugged her shoulders and added, "Some of the others are younger and perhaps richer, but she says it is to be you. Well, my son, what do you propose?"

'I agreed to maintain her in comfort as long as I lived. I gave her a duly executed bond to that effect. If there were any children she would have custody of them, and there would be one additional servant for each child.

'The old lady accepted the arrangement, and that was how I came to be tied to the Maid of Amritsar. Soon after the arrangement had been made her mother died.'

Ranjit took another drink and sat down silently. He

was exhausted now. During the telling of his story, up to the point of his success with Askari, he had been vital and radiant. He was reliving the pathway to his sensational success. But at the close of his story the light in his eyes went out, and his gesturing hands fell limply in his lap. His body, too, slouched in his chair. I wanted him to answer more of the questions in my mind, but it was obvious that he could not continue. I noticed that there were deep dark shadows under his eyes, and his usually redolent complexion had turned a sort of muddy grey. After resting quietly for a few moments he looked at me and said with weary concern:

'My dear Gyan, you look terribly exhausted and worried. Your face appears puzzled — I don't know why it should. I have said so much. The rest follows — you can bridge the gap between what I have told you and what I now am, can't you? Anyway, it's almost time for my morning yoga. We'll talk again.' He got up and flung on his coat. A moment later I heard his fast Bentley booming out of my courtyard.

Narinjan's early morning visits became again a daily occurrence. On my return from my morning ride I would see him pacing up and down the lawn, or, if the dew was heavy, along one of the gravel pathways. His head would be slightly bent and his face looked as though it were pushing hard against an intangible barrier. When he came up to me, his features would relax and break into a smile. But his words, which were invariably in the following strain, would be disconcerting:

'Sir, you have returned just when I was seeing the most gloomy visions. That means you will save the situation for me.'

I had been out riding and enjoying the morning air, and perhaps thinking piecemeal of some book I had been reading, or of Ranjit or Askari and Pushpa, and I did not see how I was going to save whatever situation he had in mind.

'But what has happened? Is Jupiter troubling you again, Narinjan Das? Or is there a worse star still in the sky now!' I would ask with a smile.

'No, sir. No, sir. No worse star, thank God; and thank God also for your power over Jupiter!' — and he would look at me as I were truly Shiva or some great guru endowed with supernatural powers. I could not help feeling rather insignificant and precarious when Narinjan talked like this.

'Sir, terrible things are going to happen unless you save me!' he went on, his face serious and pleading.

'Come, come, Narinjan! Things have gone fairly well for so many years. Good and bad things happen. We must face them and go on,' I said, putting my hand on his shoulder and speaking in a tone of gentle reasonableness.

That would calm him. His face would resume its repose, but as he left he would be muttering again about the wickedness of Jupiter.

Pushpa also called at the house each day. Her ostensible reason for these regular visits was that she was soon to return to Delhi to read for a Master's degree, and there were books in my library which she simply had to devour before she left Amritsar. But her behaviour was markedly agitated. Some excuse or another would often bring her to the house twice a day. Then, after a week or so, she abandoned all pretext of excuses for her visits and spent much of each day in my sitting-room reading or talking to me.

She talked, as before, of the romance of travel and also of her desire to be brought to rest somewhere, held down under a great canopy of happiness; laughter and life springing up around her. But to those daydreams she added another subject. At first she broached it very tentatively and by inferences. She asked me what was missing in her scheme of things; what, in fact, could hold her. She searched my face for an answer, making it more difficult for me to reply. Besides, had I come to rest? What held me?

I swallowed my own sense of frustration and said, 'Do not let the urges of youth carry you so swiftly forward that you sweep past that which would bring you what you want. The tenderness of life can be easily lost.' I stopped short, before my own yearnings poured forth. But Pushpa was still scanning my face, and she read my thoughts.

'Yes,' she said, her voice tremulous. 'Deeply ... deeply, I want tenderness and warmth and affection, but ... ' Now she stopped for a few moments. Then she went on, saying that she was on the threshold of deep love. Already, she said, she was in love in a way, but it was an altogether impossible passion. Her eyes flashed as she spoke, and then changed quickly to a softer glow.

At times she was on the verge of tears, and gently I would tell her that things would work out for her. She was beautiful and intelligent, and had no reason to doubt her capacity to find love and to explore and hold it. But she would shake her head and look very sorrowful. 'It is all so difficult. What earthly use are beauty, youth, and intelligence when the things one craves for are deep admiration and tenderness from someone who is mature and understands life? I am too young, too awkward and foolish to be able to attract the feelings of anyone who is worth while. And will I ever succeed? Much more likely that I will fumble and hurt myself, and so for ever lose what I really want.' Her fears and premonitions were obviously stimulated by her youthful sensitivity, but I was unable to reassure her. Nevertheless I wanted to help her, and I spent more time with her each day, thinking

141

that some useful course would suggest itself as we talked.

These daily meetings with Pushpa stimulated me. My morning rides became longer, I worked with greater zest in my library, and again talked with Narinjan — so far as his astrological preoccupations would permit — about business affairs. I felt myself sloughing off a heavy dull sense of frustration. But still there were bristling, intractable questions in my mind. Would I, for example, be willing to fight for the best of life that appeared to be coming my way? Would I go after Pushpa? Frankly, I did not know.

Karam took to visiting me again. He looked discontented. His heavy energy brooded sombrely in his eyes. Soon he revealed the nature of his grievance against life. Sita no longer had much use for him. They had a grandson, and she preferred the infant's company to that of her husband. Again I found myself trying to be the comforter, but the role irked me; particularly with Karam. I was out of sympathy with him. In youth, his animal reserves had been a potential of power which, one could hope, would be poured into life; but now at forty he was becoming a snarling old tiger, and I had no desire to play at being his tamer. I did not encourage his visits, and tactfully suggested that he should go off to his village and work in the fields or spend the day shooting. He took it good-naturedly; indeed he seemed to have become too insensitive and confused to take offence at anything. But his reply surprised me.

'But, Gyan, though you are a few years older than I, it is you who look as if you could go and work in the fields. What has come over you? You are bronzed and agile — more like twenty-five than about forty!'

I felt that way, but had not realized that my energy was so apparent. To what purpose was this flowering of mine, I wondered again?

A few days before Pushpa was to leave for Delhi, I planned an evening for her, but Narinjan was not well, and the party had to be abandoned at the last minute. It was then that I realized what a deep emotional turmoil was stirring within me. My disappointment at not having Pushpa with me swept me off my balance. I had expected so much of that evening. Arbitrarily I decided that Narinjan, who had visited me that very morning, was probably not really ill. It was just that Pushpa had changed her mind and refused to come. I tried to regain some calm, but it was no use. I had let myself get into a state from which relief would have to come from the outside. I told myself that for too long I had dammed up my own inner life. I had assumed that I could live by reading, riding, and helping others to restore their own inner order. That was miserably inadequate and unreal. I wanted a close tangible reality. What was I to do? The thought of Askari suddenly possessed me. I had often wondered about her. But I had no way of satisfying my curiosity. I did not know how I could reach her. Ranjit had driven me rapidly to and from her house in the old city and I would not know how to direct anyone to it.

Now an overwhelming desire to see her again took hold of me, only to be mocked by the bitter frustration of not knowing how to arrange a meeting. But a solution presented itself. Rustam had arrived for the cancelled party. When my perplexed gaze met his smiling dark eyes I realized in a flash that he, close acquaintance as he was of Ranjit's, would know about her. I broached the matter.

'By the way, my friend Ranjit has an acquaintance known as the Maid of Amritsar. Since he is not here I have been wondering whether I should not inquire after her well-being — in fact, he wished me to do so.'

Rustam smiled knowingly and nodded. 'The Maid of

Amritsar is a great patron of music. Often I sing to her. And sometimes she sings too — in her own way she is a great artist, sir. But she needs to be deeply stirred. Lately she has seemed to languish.'

Was he egging me on, giving me an opening? My mind was too set in its plan to consider and to hold back. I seized the proffered chance.

'That is unfortunate, Rustam — perhaps you should sing to her more often. It is still early tonight,' and I looked at him questioningly, restraining the expression of my most pressing thoughts.

But we understood each other. It was agreed that we should go to her house. We got into my sedate sedan, Rustam by my side to direct me, and his musicians piled in behind. The door of her house opened when he had tapped lightly and called his name. I entered after Rustam and the musicians.

We all awaited Askari in her softly furnished sitting-room. I heard the sound of nimble feet and the musical tinkle of her bracelets. As she entered the room she was saying, 'Rustam, you young dark traitor! Why have you betrayed me for so long? No word of you, no songs to gladden me. What have I done to deserve this?' Then she added, as her contribution to a conciliation, 'If you are very good tonight, you may even listen to me sing.'

Rustam bowed, his folded hands raised to his forehead. He remained silent, and realizing that there was some-thing unusual in the situation she looked around the gathering. On seeing me, her eyes changed from their quick glancing playfulness to a formal politeness — it was wonderfully done, without even a shadow of baffled surprise — and she greeted me, saying, 'Ah, the friend is here again! You are very welcome, sir, and do not take any note of my banter with this young old friend who sings to me. In truth, he has not been neglecting me, but sometimes if a single night goes by without the sound of exquisite music in this room I feel as though a black heavy aeon had rolled over me. No, it is not Rustam who neglects me; but these days life is rarely as kind as tonight when you are here!'

Refreshments were brought in and the musicians seated themselves with their instruments. Askari gave me the place of honour by her side. The music started. Rustam sang with a new emphasis in her presence, as if he felt that he had to win reaffirmation of his mastery or lose his reputation in the world of music. Askari beat time and nodded quickly when he achieved a special trill or a throaty flavour. She seemed entirely absorbed in the music, and I had a feeling of insignificance. My burning desire to see her again had brought me here; but, though she had greeted me graciously, she no longer seemed even aware of my presence. I felt utterly excluded and wished I had not come. I wanted to leave immediately, but a dull coldness congealed my limbs as I realized that if I were to get up and leave even my absence would be unnoticed.

As the last note of the song died on Rustam's lips, Askari took it up: a long low note on which she stayed to give the musicians a moment to adjust themselves to her key. Then she started to sing Rustam's song, her face raised and her hands gesturing delicately, oblivious of all in the room. I felt more excluded than ever. It was then that Askari turned and sang her next song to me. Her eyes looked straight into mine, and her words asked directly for my attention. Immediately I was electrified into life again. I do not remember what she sang — the words were unimportant. It was the fact that she was enmeshing me with the power of her eyes and the movement of her lips.

She sang the whole song to me, and when she stopped her face was masklike. She had said her say. She was

145

content to await the results. From her there was not even a smile now, nor a look seeking approval.

Her addressing the song to me was probably also a signal to Rustam and the musicians, who withdrew leaving the room to Askari and myself. I told her that her singing had brought me to life.

She shook her head. 'No,' she said. 'You did not need to be brought to life. Your aliveness brought you here. If I had been able I would have long since come to see you. So what does it matter that it is you, you who have taken the step?'

I was strangely elated and yet calmed by her words. I stroked the soft blue cushion on which I sat. It felt as if I belonged. 'Yes, I came because I had to. I had to live. Too long I have been existing with other people's emotions — against my will becoming involved in interfering in their lives.' As I said this she blinked quickly. I was baffled, and tried to change the current of my conversation. 'But why recount this? It is important only that I have come because I had to.'

Askari got up swiftly and extended her hand towards me, beckoning me to rise also. Then looking at me ingenuously she said, 'I do not know if you know. I am bound to your friend. I do not think he wishes me to be tied to him any longer, but he has not released me. You remember when you came here last I called you wise. Now I know you are.'

I did not understand her. Again she made me feel isolated. I sighed, and shaking my head I said, 'But I am not behaving wisely. Then how can you have meant it?'

Askari took a step towards me, and taking my hand in hers, she said gently, seemingly ignoring my last remarks, 'You see, I am bound to this narrow space. I am imprisoned. Your friend shuts his mind to this. You do not. You know that I am bound with many chains. If

I am to escape, I must break them all.' She smiled at me
sadly, and nodded as if to imply that she had not really
brushed aside what I had said.

I stood silently. It had been so simple to get Rustam to
bring me to see Askari. In spite of the complexity of her
situation I felt drawn to her and was glad I had come.
But was it right for me to attempt to open the barred
doors that held her? I did not know.

She knew she had silenced me with the weight of this
question, to which she herself had the answer. 'Do not
let it weigh on your mind, Gyan. The chains can be
broken by no one but myself. Telling you has helped me
greatly, and in my freedom — if it ever comes — I will
never forget it. You do not know, but you are the first
person — other than musicians, singers, poets, my women
friends, and, of course, Ranjit — to visit me in the last
four years. To me this is the beginning of a new incarna-
tion. What it will turn out to be I do not yet know.'

She was silent now. After a few minutes she said with
perfect courtliness, 'Sir, it is late now, and you must be
tired. May my women bring you some refreshment before
you leave?'

'Thank you, Askari. It is very kind of you, but I must
go.' And I raised my hands to bid her farewell.

'It is I who thank you,' she said, raising her hands. 'I,
too, love freedom, and you have brought the beginning of
it to me. Farewell.'

I drove home. My body experienced the expected
jolting movement of the car over the uneven paving and
I recognized the familiar landmarks along the streets. I
entered my courtyard, and my chauffeur came and took the
car to the garage which used to be the elephant stable. I
climbed the stairs to my rooms on the second floor. I
changed for the night and lay down.

I was deliberate about each action, aware of each of

147

them, hoping in this way to induce my mind to become aware also of the meaning of the evening I had just spent with Askari. It had moved from point to point and I could retrace now the course of its development. But the sum total of the evening I had not understood, however wise Askari might think I was.

The next morning I was up early, riding hard along the fields to the north of Amritsar. The harvest had been gathered in and the pale golden rice stubs scratched the heavy air over the dry fields. It was a transitional season — the richness of one harvest had been gathered in and this was the period of waiting before the ploughing for the next season's sowing. Then if the seed fell in the furrows, and the fertilizing rain came in time, another harvest would be in preparation. At times one is susceptible to the influence of such analogies; and that morning I felt the dry bristles within me would soon give place to seed that would germinate and sprout in sweetness. As I galloped and felt the freshness of the morning air brush past it was as if some light touch from the previous night was uprooting the bristles in my heart.

Late that morning, and for the first time in months, I opened the door to the secret room. Though the dark-winged light of the night still lurked in this western room, its colours glowed with the freshness of a moth's wings emerging from a chrysalis. I stood in the centre of the room's perfectly composed beauty. The paintings on the walls continued to depict authentically a life of activity and love in unbroken rhythm and beauty. Love, rhythm, and activity — all had eluded me. People regarded me as specially favoured by life — I had education, wealth, I had married and raised two sons; and my friends looked up to me and sought my counsel. But, standing in this room and looking at the paintings, I melted again into a deep agony. I recollected my own youth and my life with Basanti, and how all that period I had felt excluded from this room and the warmth and beauty it represented. I stood till the sorrow seemed to mellow, and through it I felt a gentle stirring within. It was as if over the inner liquidity of sadness a light ripple was playing.

The day before Pushpa was to leave for Delhi she paid her last visit to me. I well remember her rust and gold sari — colours incongruous with her youthful years but pronouncing the mellowness and depth of her feeling. She brought me a beautiful bookmark on which she had painted two heads facing each other. Between them was a spectrum of light in which the colours nearest each face were full and even heavy, but as the two ends of the spectrum approached the centre the colours became increasingly bright and translucent — yellow, gold, mauve, and blue. She said shyly, as I was looking at it, 'They must both move to the centre.'

Looking at the beautiful colours of the bookmark, I had an impulse. 'Come.' And taking her arm I led her to the door of the tapestried room and threw it open. The breeze coming in through the open door to the balcony blew her sari close to her, and she stood there for a few moments, her body taut and excited, responding to the loveliness of the room. We entered and stood in its centre. Silently she turned around to look at the room. She seemed to take it in, and then stood very still. Then with a quick movement she came and stood pressing against me. Gently she put her arms around my neck, and looking at me with her large eyes, now very brightly lit, she nodded her head slowly and said, 'Gyan, this *is* the centre.'

I held her close and felt the bristly dryness in my heart change to a warm throbbing quickening towards the beautiful girl in my arms. Perhaps this *was* the new beginning that had been surging up in me. In utter

149

happiness I gently raised her face to mine. But there was fear, almost panic, in her eyes. Quickly I adjusted the expression of my face, and pressing her closer I looked at her reassuringly. After a moment she sighed softly and her eyes warmed to mine again. But though the fear had fluttered away the look that remained was not the magic trustfulness of a moment ago, when she had said 'this is the centre'. I still held her gently, but I was disturbed by the change in her and I felt my body become very still. My eyes wandered from the room to the balcony. The late afternoon light was redolent on the old brickwork. Confusedly, I felt that the moment of magic had returned. I looked at Pushpa, but her eyes now were far away, as if she had lost something and had no hope of finding it. A strange urge possessed me to help her find it. Somehow I knew that this thing she seemed to have lost was vitally connected with the two of us.

Gently I released her and led her to the balcony. The light of the setting sun was sprinkling its profuse deep gold on the trees, the roofs, and the few tall white buildings of the old city. The evening was offering us its largesse extending beyond the city to the soft powdery light that hung over the dun and blue fields till it reached the source of the golden light itself. Pushpa was moved by the beauty of the scene. We watched the light fade, not breaking the spell by uttering even a single word. Desperately I hoped that in the silence she had become again the girl who had shyly brought me the faces coming together.

We returned to my study. Could we not immediately reach the centre of her picture? Need she leave Amritsar just to take a university examination at Delhi? No, these were wild thoughts, but she could surely hasten back from Delhi. If this was the centre there was no reason why she should not quickly return to it, I told myself. Even if she now doubted that this was the centre, let her go for a while to give her doubts time to resolve themselves one way or the other. But could I put all this to her? What could I offer her? Too long, ever since Jennifer Trip inevitably had left me to join her parents at Calcutta, the flow of womanly tenderness had skirted past me. I did not know quite why I had gone the previous evening to Askari's house; and I did not understand what had occurred there. Was it not again the sudden unexpected turning away from me of the promised sweetness of the tenderness of a beautiful woman? Something now held me back from making an offer to Pushpa; but somehow I had to trammel her departing step, to revive her need for the open place that craved to be filled with her presence.

My throat hoarse with emotion and apprehension, I said, 'Pushpa, even the books in this room will miss you. Who will read them now? Take some with you and bring them back when you return to Amritsar. Perhaps your stay in Delhi will not be as long as we think.'

She gave me a quick glance and smiled remotely, as if saying 'You do not understand!' Her words confirmed my apprehension. 'Thank you, Gyan. I will take some books, but' — and she shook her head — 'the books alone cannot hold me here. And yet, I could be kept from leaving. I could stay here. I could forget my dreams of travel, meeting wonderful people, and seeing the world. But somehow it seems this is not to be. I feel I have to go to Delhi now. But the books: I will take them.'

She stood before me, very beautiful, poised, but remote. Could I not say something, could I not cry out so that she would hear my inner yearning? She turned towards the bookshelves and chose a handful of books. Silently we went down the stairs together.

I felt I was surrendering to the wide world all the warmth and beauty that had for a short time so un-expectedly entered the old house. I was to be left with

151

nothing: only the ancient mansion, the high-ceilinged rooms filled already with the empty darkness of the approaching night.

The next day Narinjan came a little later than usual. He had been at the railway station seeing Pushpa off to Delhi. He looked very sad, and expressed grave forebodings about his child's departure. Though my own sorrow clogged all my impulses I tried to cheer him. 'But, my dear Narinjan, she has been to Delhi before. Why do you let your sadness cast a gloom on the future?'

Narinjan sighed. Then his eyes suddenly glowed as if illumined by some sure truth. His face was still curiously youthful. What was it that sustained him? Slow and oracular his voice came. 'Good sir, I do not wish to cast a shadow on the future, but the dark umbra of Jupiter is very strong at this time, stronger than ever before.'

I looked at him. It was his faith in the occult — to him the essence of wisdom — that sustained him, for his whole face beamed as he made his sombre announcement.

Silently we continued to pace the lawn. Then Narinjan said, 'Your friend, Ranjit, was on the train to Delhi. I requested him to keep a kindly eye on Pushpa.'

This was amazing news. Ranjit was to have left Amritsar several weeks ago for his long trip to the east and into Tibet.

'Ranjit? Are you sure? He left weeks ago for Bengal!' I exclaimed.

Narinjan replied phlegmatically, 'I remember one evening here he was talking of his trip. I asked him about it at the railway station. He told me he had had to postpone it but hoped now to go on to Bengal.'

It was strange that Ranjit should have remained on quietly at Amritsar, not informing me of his changed plans. And was it just coincidental that he was travelling to Delhi today?

For the next few weeks I continued riding hard in the mornings and working at my business affairs. I had neglected my landed property. There was a growing demand now for sugar cane, cotton, and tobacco. I decided to stimulate the cultivation of these crops on my lands. Several times I rode out across the thirty miles or so to my farms and made arrangements for the best varieties of seed for the crops which I wished to encourage. I was amazed to find how unpopular my managers had allowed themselves to become. Indeed, the ignominy which was universally heaped on them was well deserved. They had been merciless in their exactions from the peasants, and I was not informed of the seasonal droughts or pests which made it difficult for all the peasants to pay their rents, or of the difficulties on the holdings of individual farmers caused by such occurrences as illnesses, or death of plough cattle, or a well having fallen in.

To encourage the new crops I reduced rents on all lands which should be brought under their cultivation, and I dismissed two of the most notorious managers. It turned out that these changes were timely, for soon after, when in the district of Amritsar there was a serious wave of unrest among the farmers, my men — who had received better terms than the others were demanding — went on expanding the cultivation of sugar cane and cotton. Not that I had acted with a view to forestalling the possibility of discontent on my farms. I had been upset at the conditions I had seen and had done what seemed to be necessary to effect some improvements.

All through those months of going out to the villages and supervising my affairs, I lived for the time when Pushpa would return. In spite of my forebodings, and the events in my study before she left, I told myself that her absence was only temporary; and this thought spurred me on in my daily activities. Nevertheless, periodically

153

an arresting fear would petrify me: Pushpa was young and beautiful; what if she should suddenly decide to marry someone in Delhi? I would never see her again, and this period of hard work would not lead to the sweet recompense of her daily presence in these empty rooms. At times I would closely question Narinjan about the progress she was making at college. Good fellow that he was, he took my interest to be fatherly and would embark on a discussion of what she should do next.

'Sir, in your wisdom, you must help me find a suitable husband for her. I do not understand her very well; but in recent years she has been like a child in your home — reading, talking to you, and learning the ways of this gracious house. You must advise me where to seek a husband who will bring her some happiness.'

Madly I wished that I were my own son and that age was not so blatant a barrier as to stop me from saying, 'Narinjan, do not worry. I myself will marry her and she can live in this old house which has become so accustomed to her presence.' But that absurd sentence never crossed my lips, and instead I could only say, 'There is still time to decide, Narinjan. Let her return and we will see her mood. Then we will find her a husband.' Little did he realize how much anguish filled me as I formed these empty words and suppressed my feelings for his daughter.

The months passed. Summer was approaching, and soon Puspha would be due to return. It was in that period that Rustam the singer came and saw me a few times. I told him that I was looking forward to arranging an evening when he could sing to my friends. We talked about the songs he had been learning, but he had something else on his mind. He arrived at the house early one morning and met me in the garden.

'Sir,' he said, bowing so low that I expected at least a request for some kind of assistance, 'I have been wanting

154

to tell you something.' He stopped, and I looked at him wonderingly.

'Then do tell me, Rustam,' I put in, as encouragingly as I could.

'Great sir, it is really only a message from the Maid of Amritsar. She has heard nothing from your friend for long months and she wonders if she could see you. She thinks you might be able to inform her of what has happened.'

Rustam was very likeable and I had no wish to offend him, but I had no desire to see Askari. What could I tell her about Ranjit? Besides, I had no wish to experience anything like the strangely baffling evening when I had visited her. And, what made me most disinclined to see her, Pushpa would soon be returning to Amritsar. I had no desire to give Askari the impression that I wished to renew my acquaintance with her.

Rustam, whose eyes were remarkably expressive, looked as though he divined my hesitation, so that I felt I could speak freely.

'Let me explain the matter to you, Rustam, for I do not wish the Maid to think you are lacking in the power to be persuasive or to dispose people to meet your requests. But, you see, I do not know where our friend Ranjit is. I could tell her nothing. Tell her that. Tell her also not to worry. If he is in Tibet it is a remote area and it is not surprising that he has not written. You might also tell her, if you wish, that I think Ranjit is an honourable man, and he will, when he can, get in touch with her.'

Rustam left, pleased it seemed to me to be entrusted as my spokesman.

It was fortunate that Rustam came when he did, for I could then truthfully say I knew nothing as to Ranjit's whereabouts. A few days later, without warning, his Bentley — a vintage model by now — boomed into my courtyard. Ranjit came to my study upstairs. He was leaner, keener-eyed, and restless. He was even more carefully dressed in his casual sporting manner; but somehow he did not look as if he had been braving it out in the distant high steppes of Tibet. His look was that of a man of intensified sophistication, and not of someone who had been scanning the wide spaces and the heavens.

'My dear Gyan, this is a pleasure I could not deny myself after all these months! And besides, I may be here only briefly.' He said this quickly, too quickly to mean it more than as a pleasant introduction to something else.

I was curious, and tried to draw him out. 'Do tell me, Ranjit, of your travels to Bengal, Tibet, and elsewhere. You have been away a long time and there must be much to tell.'

'Yes, yes, there is much to tell. And I have travelled. I never got to Tibet. In fact, I never even reached Bengal or Assam. I did a little shooting in Central India — much closer afield — and I met many interesting people. Indeed, strange as it may seem, I have become rather interested in politics. I have spent a good deal of time with the leaders of all shades of thought in Delhi. One hardly realizes it, in this pleasant backwater, but things are moving very fast in world affairs. I wanted to go to Europe, but after Munich my friends regarded that as unwise and a waste of time and effort. But I am rambling from one thing to another and not telling you anything.' Though he as good as admitted that he was parrying my questions there was a wary look in his eyes. He was jumping rather than rambling. In fact, he was cleverly dodging what he really wished to tell me, but I had a feeling that it was best to let him talk in his own way. He would become exhausted by the strain of trying to contain himself, and then he might well blurt out what he most wanted to say.

I smiled at him and merely chipped in, 'Yes, you are hurrying yourself — unnecessarily so. Let us have a drink and relax. Then you can tell me at your leisure all that has been happening to you.'

We went down to a sitting-room on the ground floor and I poured out a couple of drinks. Handing Ranjit his glass, and thinking of Pushpa's impending return and my talks with Rustam, I said casually, 'Would it not be fun to have Rustam sing for us and our friends here? Let us fix something one of these evenings, Ranjit.'

I expected him to hail this suggestion with enthusiasm. To my surprise Ranjit's hand shook, spilling some of his drink, and his voice sounded disturbed as he said, 'What? Rustam? No. No, Gyan, let's not arrange any such thing.'

Seeing my surprise he added — but obviously quite im-promptu and meaninglessly, 'I know of much better singers. If we are to arrange anything we will call them in.'

My train of thought quite naturally led me to my next inquiry, even though I had an inkling that it would not serve to calm Ranjit.

'What news of the beautiful Askari?' I asked, effecting a tone of friendly inquiry.

Ranjit tried to look unconcerned, but he was on his guard. I could see that he felt I was probing him. His eyes gave him away as he said casually, 'I don't really know. Have you kept in touch with her at all?' Now he felt he had turned it neatly on me, and his eyes brightened as he sipped his drink.

My memory played a strange trick. I suppose it was not so much that I wished to conceal it from him, but rather because I wanted to exorcize the memory of it from my mind, that I had no recollection of my visit to Askari.

157

But Rustam's visits were fresh in my memory. I told him about them and of Askari's inquiry.

Ranjit's face flushed. For a moment I thought there was going to be an outburst from him, but he struggled and gained mastery of himself. Again speaking with deliberate casualness, he said in his deep warm voice, 'Gyan, that is one of the things I really wanted to tell you. I mean about Askari. I think I told you how I came to be bound to her. Well, my responsibility is really limited to a financial settlement. I undertook no more than that. Whatever happens I will stand by that undertaking,' he said, looking determinedly virtuous, forgetting that I knew of the written deed to that effect and that he therefore had scarcely any alternative. I felt sorry for him. I knew him so well. I wanted to go up to him and put my hand on his shoulder and tell him he could say what he really had in mind. But his expression and manner were forbidding, and I had to let him continue in his own way.

'Yes, I will stand by my undertaking, but that is all. I do not want to see her again. I have other things on hand now — politics, in particular. You will understand why I am not keen to have Rustam around. He is a friend of hers. She has no reason to know where I am; my bank will continue to pay her allowance, and I will in due time inform her that that is the only contact with me she is entitled to retain.' He stopped and sipped his drink. The determined look on his face collapsed. I could see that he was wounded by the harsh stridency of his own words. He looked at me quizzically, wondering what my response would be.

Ranjit evidently expected me to react strongly to his statement. But why should I comment at all on his decision? When he took me to call on Askari it was perfectly obvious that he wanted to have nothing more to do with the girl. In any case I disliked interfering in the affairs of other people. So I made no comment, and reverted to another topic.

'Tell me more about your travels and your political work. You have beaten your sword into a pen or a rostrum. It is a fascinating change, Ranjit. Tell me about it.' The ingenuousness of my inquiry must have been clear from my face and voice.

To my surprise my words upset Ranjit. He shifted uneasily and his face flushed testily. Again he struggled with himself, but less successfully than before. Rather rudely, he said, to the accompaniment of a forced laugh, 'It really would not interest you, Gyan. It's much too complicated.'

I shrugged my shoulders and offered him another drink. There was an uncomfortable silence. I did not feel I could raise another topic with him. He realized that he had silenced me and tried to make amends.

'Gyan, the truth is I have been caught up in so many things that I find it difficult to adjust myself to life in Amritsar any longer. I don't know that I shall come here again for a long time. I am interested in international affairs, and the war that now looks very likely will create new problems for our political leaders. I could easily find myself involved in some of the situations which will develop. Amritsar now upsets me with its unchanging scene. God knows, I do not want to see it again for a long time. I might travel to Ceylon in the near future for a short visit.'

I listened to all this, feeling that there was a link somewhere in the new developments which Ranjit was concealing. Not that it was strange that he should have become interested in politics and international affairs, but it was quite unnecessary that these new interests should be accompanied by a complete break with his past and his usual surroundings. I would feel genuinely sorry to see him leave Amritsar for good. Though I could not take them entirely seriously I had to accept his statements.

'I am sorry you will be leaving, Ranjit. In fact, I was hoping that after your trip to Tibet you would spend some time here, and among other things nurture our friendship.' Then thinking again of Pushpa's impending arrival, I added, 'but let us, in any case, arrange a pleasant evening together. I would like to see Karam and his wife, and some of our other friends, with us again.'

Ranjit made an effort to enter into the spirit of my suggestion. 'Good,' he said. 'Tell me if you would like me to arrange for the musicians.'

Allowing for two or three days after Pushpa's arrival I suggested a date in about two weeks' time, which suited Ranjit very well. He was more relaxed now, and I was thankful that we had got away from inquiring into each other's affairs. We had another drink, and before he left Ranjit was himself again, saying that he had things to attend to on his lands but that he would soon pay me another visit.

Those few days before Pushpa's arrival I filled to the brink with activity. I rode out to my farms several times, and even enjoyed the ride back in the hot afternoons. On the way there and back I would break journey under a grove of great fig and mango trees about a furlong from the main track to my village. It was a beautiful, quiet, and undisturbed place, and without any apparent geological reason there was a pleasant undulation of the land — which had prompted someone, perhaps a century or two before, to build a spacious tank in one of the hollows of the land. The grove was a charming place. The only inhabitant appeared to be a tall, wise-looking, and silent recluse. He had observed my visits but took no notice of me.

While awaiting Pushpa's return, often I visited the

tapestried room. Each time it seemed to glow more wonderfully, reflecting my own heightening emotion as Pushpa's day of arrival approached.

The day came. I knew from Narinjan, who seemed to me unnecessarily agitated, that her train was to arrive at about eleven in the morning. Should I go with him to receive her? My own urge was to rush to the station half an hour in advance in case the train should be early. But I argued myself into a more restrained frame of mind. It would not do for me to go; it would look too obvious, and my emotional state was such that I would probably display such a degree of pleasure on seeing her, that it might embarrass the girl. It was best, then, that I should wait at home. Narinjan would certainly call on me soon after her arrival to tell me how she was, and it was not unlikely that she would accompany him.

Each minute of that morning seemed like an eternity. I paced up and down in the garden, looked at my books, climbed to the little tower rooms, and scanned the hot horizon on all sides. But nothing lessened my agitation. I felt my nerves fraying in spite of my effort to recall myself to an orderly calm.

Somewhat earlier than I had expected I heard the familiar sound of Narinjan's old car. I looked at my watch. It was just about eleven. So, after all, the train had been early, and at least in a measure my tedious and agonized waiting had been abreviated. The car stopped, and I heard Narinjan's footsteps, unusually accelerated. He ran up the front stairs and across the veranda.

Pale, and his face reduced to the helplessness of a child's, he rushed into my room holding a sheet of paper in his outstretched hand. His voice was quivering with emotion as he cried out, 'Sir, please read this terrible news!' He sat down, his head buried in his hands, as I read the telegram.

Dearest Father,

Yesterday Ranjit and I were married. Today we leave for Ceylon. Hoping to return in a month. Will write regularly. Soon we will meet.

Love,

PUSHPA

The message had been sent from Delhi.

I looked at poor Narinjan, reduced to sobs in the arm-chair near me. My own reaction amazed me. It was as though I had been expecting this to happen. Like a flood there came back to me the sudden changes in Pushpa's mood before she had left for Delhi. How foolish, then, to have permitted myself to live only for her return! My keyed up expectancy, and the first shock of disappointment on reading the telegram, both seemed ridiculous now.

But there was no time to deplore my own foolishness. The immediate problem was Niranjan. I found myself searching desperately for some way to show him the best aspects of the situation.

'Narinjan, do not grieve so. At least it is someone you know. He is a friend of mine — a likeable, honourable man, on the whole. Of course, he is not young, but then ...' I broke off, realizing that I was going to say that neither was I — 'she knows that, and it has made no difference to her. It is much better than if she had run away with some young man of unknown background, or perhaps of small means. Come, let us face it bravely.'

Narinjan kept sobbing. I put my hand on his shoulder. His sobs suffocated in his necessity for breath. Then he sighed deeply several times.

'Why all this ... all this secrecy? He was here ... I saw him at your home. Could they not have told us and asked for our blessing? Am I not her father, and you even more than a father — her benefactor, her father's protector?

162

What behaviour is this!' Again he sighed and moaned.

Already I had succeeded in convincing myself that it was not as bad as he was making out.

'Narinjan, perhaps to some extent we were to blame. Obviously neither of them felt able to take the risk of telling us. They felt we might not approve and would try to stop them. Would we have?'

Stupidly I half hoped that he would say that of course he would have stopped them. But his words immediately stung me back to reality. 'No sir, no. Why should I, if that was her wish? He is, after all, your friend. That is enough for me.'

'Well then, Narinjan, if you would not have objected let us forgive them for having done what we would have approved. They have anticipated our approval — that is all.'

For a moment he was silent. Put that way the situation was much more acceptable than I had imagined. In the attempt to soothe Narinjan I had fully convinced myself. Even he seemed to feel that what I had said was reasonable. But after a respite, again he moaned.

'I knew the stars were fiercely against me this season. Did I not tell you so, great sir? But your words are also right. Even if you have uttered them only to comfort me, I accept them from you. We must ask them to come here, back to their home.'

This I did not want. The thought of their returning to Amritsar together brought my own anguish sweeping back over me. But I felt I had to continue in my role of comforter.

'Of course they will return. You will visit them also. It will be something to look forward to,' I said without conviction, as I struggled to control my own feelings. Narinjan noticed the change in my voice and groaned again.

'If it is God's will that I am to lose her, what can I do? And Sulochna her sister thought so highly of her!'

Narinjan spent the rest of the day at my house. At the end of the evening we were both exhausted. But, in spite of my emotional stress, I fell asleep early.

The next morning I awoke with an extreme emptiness gnawing at my stomach. Alone now, the full impact of what had occurred overwhelmed me. Confusedly, I had been building my life on Pushpa's return — I had never faced squarely the question as to what I would do when she returned. Timidly I had assumed that one step at a time was the best way of proceeding; then, in time, our lives would converge and we would spend our days together. What a fool I had been! — hesitating because of my age, and pusillanimously drawing out the process as though I had for ever! Naturally, then, I had lost to a man of sense who had grasped the opportunity offered by no more than two or three meetings with the girl.

My mind seethed with vivid pictures of the girl's warmth, her sparkle, her gracefulness, her expressive eyes, her extraordinary beauty — and I was numbed by the cruel lash of fate and foolishness that had deprived me of so much. Never had I experienced the continuous flow of womanly love, and now that I could have slaked my consuming thirst I had failed utterly. As to Ranjit, I could not hold anything against him. He had been wise where I had been foolish. He, too, had needed affection. I had seen it in his eyes, and now it was all to be made up to him. No, I felt no animosity towards him — only the feeling that he was so much more fortunate than I.

As the days passed the flatness of life affected me like a consumption. I felt it eating me up. My skin — so long fresh and taut — started to gather under my eyes and on the back of my hands. My hair became sparser and rapidly greyed. I welcomed these changes. I wanted to achieve a complete metamorphosis. Sometimes, to assure myself that I was in fact incapable of reacting to those things that had deeply stirred me in the past, I would enter the tapestried room. And, as I expected, it no longer meant anything to me. The room was more remote than the pictures on its walls. In those the artist had left something which still stirred me, but the room itself might have been a scene out of the life of the mummied pharaohs.

I was glad of the rapid onset of age and weakness, but Narinjan, who seemed to thrive under the evil influence of Jupiter, was worried for me. He insisted that I escape the overpowering heat of the summer by going to the hills. He took a place in the mountains and there we spent three months. We were tended by the quiet Sulochna, who, apart from the time she devoted to us, spent all her days roaming about the friendly mountains or working at her books. The care of my friends and the mountain air could not but revive me, and by the time I was beckoned down to the plains by a petition from my farmers I was well and robust again, if decidedly older in appearance.

I was able to settle the situation on my farms by dismissing another two of the old group of managers, and I returned to Amritsar refreshed by my visit to the mountains.

Narinjan was still in the hills with his daughter. I no longer saw much of Karam and Sita, who had become a querulous gossiping couple. And now my complete aloneness gave me a sense of freedom. I pottered about happily in the house, in the garden, and in the country byways while on my morning rides. It was a clear cool autumn, and the fresh leaves on the evergreens waved at me tantalizingly, challengingly, each morning as I went riding. Something in me took up the challenge. Each morning I awoke with increased vigour. I rode out to my farms and back the same day — a distance of almost fifty miles — and enjoyed my evening meal more than ever, still unexhausted.

Vigour and freedom seemed to shake me out of my old propensities. When I loved and longed for Pushpa, I had not only put her before myself in all my thoughts but I had virtually excluded all consideration of myself. Always hanging back I had ensured my own unhappiness and nothing else. Bitterly, the realization came now that it was only logical that Ranjit should have overtaken me while I wilfully lagged behind. My returning vigour and the new sense of freedom I was experiencing turned in revulsion from the vapidity of the life I had been living. I wanted now to close in on life, even if it pushed at me its pungency or acridity rather than its sweeter flavours.

It was in this mood that I sent word to Rustam to come to the house. But this time I did not wish him to sing to me. Immediately after dinner we drove to Askari's home. She entered the doorway, looking wan, the light in her eyes subdued and turbid. Her rich silk sari of pale blue, deep mauve, and gold was an incongruous remnant of a past age.

On seeing me, she became agitated and hesitated by the door. She was clearly in no mood for company, but overcame her melancholy enough to greet me calmly, adding that she was honoured we had thought of her. She asked Rustam to sing, but was listless while he performed. After the first song she signalled him with her eyes and he withdrew from the room. When he had left she turned to me. Attempting to smile, she said in her musical, expressive voice:

'You must not think me rude, Rai Sahib. I have been much perplexed of late. Your friend was in Amritsar till recently. He was seen by acquaintances of mine. You also must have seen him. But he never came to me, and now I believe he has again left — he has not been seen recently. Is it not strange and meaningless that he should imprison me, then?' There was a look of tortured speculation in her eyes, and I noticed that her eyelids were swollen.

Why should this beautiful woman remain in this absurd position any longer, I asked myself. Ranjit had married; he had told me that he considered himself responsible for Askari only financially. Why had he not told her so? Why should she be imprisoned by the unthinking neglect of Ranjit, who undoubtedly had meant to convey his position to her but had not done so because he was intoxicated by his new love, the beautiful young Pushpa? Although realizing that my motives were not concerned just with her freedom, I decided virtuously that it was right that I should release her.

'Askari, Ranjit is a friend of mine, and he is an honourable man. Soon he will tell you the position he is in now. He is bound by deed to you. He fully acknowledges that, and the term of the deed he will fulfil. But his own life now will keep him occupied — probably almost entirely — at Delhi and elsewhere. He will come to Amritsar very rarely, and when he does he will bring with him his young wife.' I stopped, and she sat up straight.

'His wife? Who says he has a wife?' she demanded, her eyes coming to life in great bursts of agitated light.

'I am his friend, Askari. I know that he is married,' I replied, trying to calm her by the softness of the tone of my voice.

Her head dropped and she covered her face with her hands. I thought she was going to collapse, but after a moment of wavering her body seemed to congeal in its stiff upright position. But the inner defiance of her being failed her, and when she lifted her face on it was the torture of bitter defeat. She struck her brow softly with her fist and said, to herself rather than to me, 'Why did

167

he not tell me? Perhaps he did not wish to hurt me. But he knew it was my dream that he should release me. And yet, he has told me nothing!' She sighed deeply.

I felt I should leave, for in her present state she would not want my company. But she looked up as if asking me to confirm her own conclusions.

'But you are released now, Askari,' I said in a definite tone to induce her to accept the fact as final. I looked at her. She was very beautiful. If only she would realize she was free, perhaps in time — and soon, I hoped — she would become aware of me.

But she shook her head, arresting my personal fantasy. 'No, you do not understand, Rai Sahib. I wanted a complicated freedom. I wished Ranjit to take me out of this house and to let me go with him wherever he went. I wanted no freedom from him — which you say he has graciously conferred on me — but freedom with him!'

She shrugged her shoulders and pondered sadly. She knew she had lost. Then again she looked up, and this time she smiled.

'Rai Sahib, you are very kind to have come here. Now at last I know my situation.' She looked at me searchingly and must have perceived my disinclination to stay. Immediately she divined my reason.

'But you thought I already knew, so you came! Now that I do know, let me invite you to visit me. Please come and see me in a few days' time. Come, and we will be two people meeting on the same footing, trying to understand life. This introduction has been painful to us both.' She got up and I left, again feeling somewhat humiliated by my visit.

For the next few days I tried to exclude all thought of Askari. She was beautiful, yes, but her life was complicated and it had given me no pleasure to be with her. Then why let my thoughts dwell on her? This was a perfectly reasonable approach and should have put my mind at ease. But it did not. I was obsessed by the recollection of her invitation to visit her again. I told myself that she had merely been polite and it was absurd to attach any special significance to her words. Convincing though this explanation was, and in spite of my rational disinclination to see her again, I realized I would need constantly to restrain myself, to argue with myself, and to find other things to do if I intended to stay away from her.

About a week later I set out one evening for her house. That I no longer went as part of the entourage of Ranjit or Rustam produced a pleasant feeling of being in control, which counterbalanced the fear that I was behaving irrationally. Arriving at her home, I entered with a sense of assurance.

It was some time before Askari appeared, giving all my doubts time to seethe again. Perhaps she would not even see me. Perhaps I had, after all, been a fool to take her invitation literally. But when she entered the room all was swept away by her very presence.

Askari was dressed in a superbly glowing red sari. Her face was radiant, and she had heightened the power of her eyes by a fine line of antimony black. Smiling with what seemed to be genuine pleasure she came up to me, saying with a quick enthusiasm, 'So, at last you have come; after many more days than I had wished to pass without you. And yet, I forbore to send Rustam to you.'

Could she have been struggling to interpret to herself her invitation to me? That thought — and had she not practically admitted this to me now? — gave me a delightful if somewhat false feeling of ascendancy, for I had not told her that I had spent restless days restraining myself from visiting her earlier.

The evening passed delightfully. We talked lightly,

only occasionally verging on the serious. She sent for her musicians, sang, and even danced for me. Then she fed me on viands prepared by a master cook, and showed her own keen and cultivated appreciation of food and wine.

But most wonderful of all was her subtle awareness of my desires. Even with the encouragement of her welcome I was too restrained and bashful to express my yearnings to her. But she sensed them, and late that evening, when a kind of rich silence fell between us, she came close to me and gently she coaxed the fires in me to come aflame for her. The night passed in a superb interlocking of passion, and deep serene slumber. In the early hours of the morning before I left she clung to me, so that I went with the feeling that she had given herself fully and wished me to come to her again.

Thereafter I visited Askari very frequently, and at last, for the first time in my life, I felt a new dimension had been added — a dimension which gave form to and shed its light on the others. My mind became acutely aware of the overtones of pain and joy in all life around me; and my senses were as if they had been peeled open.

Did I imagine that her body, when roused, became an iridescent fire? Her burning words drew me yet closer. 'It is like the hot springtime earth bursting open in me! I have never, never known this deep closeness, Gyan!' she would cry out. And then, later, she would say in a tremulous happy voice, 'Like quiet warm rain on the newly formed fruit on the trees. You make life glisten in me, Gyan.' She would draw me close again in a warm embrace.

All my life I had yearned for this closeness, but it was she who found the words to express it. I was more than content just to come to her evening after evening, and often also during the day, drawn by the power of this new experience.

Perhaps the long hours spent with her were a reckless way of living. I was expending prodigiously what strength I had built up. After a few weeks I thought I had begun to look pale and worn, though I felt buoyant and more alive than ever. It was perplexing that Niranjan, who was always solicitous about me, had not called attention to my drawn appearance. I began to wonder if he had not somehow learned of my liaison with Askari and was not saying anything for fear that I should be hurt by it. But my doubts on this score were set at rest when Karamjit visited me one afternoon. He had not come for several months. As he entered the room, he burst out, 'Gyan, what has come over you! You are looking healthier and younger even than when we were young men together.' He shrugged his heavy shoulders and lifted his hands.

'Look at me,' he went on, 'growing all over like a dozen buffaloes. And yet it isn't as if my life were one of great unhappiness or of debauch.'

I could not help smiling. If Karam only knew the cause of my looks he would have thrown a fit, I thought. I wished I could tell him. It was just what he needed to know. But I decided to be evasive. 'You are content, Karam, and I am not. That is the difference, my dear friend.'

'Well, contentment seems a pretty ugly thing,' he grumbled back. 'I wish I had your brand of discontent. But Gyan, who says you are discontented? And whatever about? You, who have everything! I suppose you are fortunate and I am not.' He gave me a sour look, as if in that way he could help his supposedly smaller fortunes to even up with mine. I let the conversation drop.

There were but two threads of pain in my life with Askari. One was my own feeling of distress that I had to visit her clandestinely and live a life of love and tenderness cut off from my day to day existence. It made me acutely

171

uneasy that when, during his morning visits, Narinjan complimented me on my good health and spirits I could not tell him what it was that was bringing me this new-found gladness.

The other strand of the pain which I felt came from Askari herself. She had told me her life story. It was one of being immured in the house in which she lived. She had come to it with her mother, as a little child, and had lived there ever since. She had hoped when she and her mother selected Ranjit that it would be the commencement of a new and wider life for her. She assured me that with me she was blissful and wanted nothing else, but I realized that she dearly wished to escape from the prison that held her.

I did what I could to fill her life. I gave her clothes, pictures, and jewellery — but all these were things with which she had been surfeited in the past. I encouraged her to spend some time each day driving about the city in a phaeton. She would see the town, and gradually she could do some shopping and walk in the park. But gently she told me that it would be like moving around Amritsar in her solitary cell. She had tried it and it had only driven her back to her house, for it provided only a false sense of normal living.

But the pain did not arrest the deepening happiness of our own lives. The intensity of our love — which I am sure it was — increased steadily, and at last I knew the meaning of an abiding, continuing flow of tenderness. It was always there, on both sides, intoxicating and sustaining each of us. A white light filled my eyes during the day, and at night it was as if the darkness became a quiet iridescence of black and green and blue before I fell into the deepest sleep of refreshment. When Askari held me my body felt as tender and sensitive as a spring plant, and yet I was filled also with the strength of an oak. I was always ready for the brisk bustling arrival of the morning, the open challenge of noon, and the quiet sad ebbing away of the day into evening. Life became a balanced beautiful movement in which time and space were both comprehended and thus brought to a kind of stillness within the perfection of movement.

In this pure joy, when there was even the slightest tugging of the painful thread, it hurt intensely. Why could we not cut this thread? I thought it over carefully. I was not young — I was approaching fifty. What if people did begin to gossip about me? It would not do my children any harm — they were a long since settled. Romesh, too, had returned from England and was a practising lawyer at Delhi. Why then should I not let Askari visit my house? Perhaps she would feel reconciled to living at her own home if she could visit me frequently. There was no reason deliberately to advertise her visits to me, but if they should come to be known, what of it? Narinjan would never gossip about me. Ranjit was gone from Amritsar, and in any event he was not the sort of man who would wish to talk about situations in which he had himself spent most of his life. Karam and Sita would fill Amritsar with the news, but I saw very little of them and there was no reason why they should find out.

I decided to take the step, and told Askari of my decision. Her face lit up, and she said the news made her feel like a bird in flight. She was soaring into the heavens. We laughed together at her joy.

It was agreed that on the first visit she would come one evening with Rustam. It would appear to my servants, and to anyone else who might hear of it, that she was a new singer who was accompanying Rustam. The camouflage hurt me, but we both agreed that it was a wise course.

It was a difficult but worthwhile evening. Askari was

full of excitement and some trepidation at breaking out of her prison walls. She liked the big house, the rooms with their mural decorations, and the great courtyard. But the solemnity of the furnishings and the silent attention of the servants awed her somewhat, and at times I noticed her startled little tremors as she shrank from something in the unfamiliar surroundings. That evening we did not go upstairs, though we were together alone late in the evening when Rustam with his musicians politely retired to the veranda to smoke and drink. Askari was tender in her embraces but a bit distracted by her new surroundings; yet neither of us let this interfere with the bliss of the evening.

A few days later we arranged another occasion. Again Rustam came with her but it was earlier than usual, and while he awaited his musicians I took Askari upstairs and showed her part of the house. Later that evening, when Rustam and his men again retreated diplomatically, I said to her, 'Come with me. There is a room I must show you.' Again I took her upstairs, and led her through my dark unlit study to the door of the tapestried room. When I turned on the light her finely developed instinct for sensuous beauty at once responded. She gasped with surprise.

'It is wonderful,' she said, turning about in the room, letting her eyes drink in the splendour around her. She darted to the Rajput paintings and studied them closely. 'They are more beautiful than life itself,' she exclaimed. 'What loveliness!'

I went up to put my arms around her. But she darted away again to look at the hangings and the paintings. She was thrilled and completely absorbed. I saw at once that having been starved all her life, she now responded with the eagerness of an intelligent child and the sensuous capacity of a beautiful awakened woman. I could have

174

walked out unnoticed — all her attention was riveted on the room.

When we came downstairs we held each other tenderly before she left, but her passion was fitful. Obviously, I thought, it would take time for her to become accustomed to the house. As she left, I asked whether the next time I should come and visit her. But she immediately shook her head in protest. No, she wanted to come to this house a hundred times.

So she came more and yet more frequently. Her animation and excitement increased. Her responses to everything quickened. She argued and discussed all that came up in conversation between us, drawing constantly on what she had seen since her release from her 'prison', as she continued to refer to her home.

One evening Narinjan was at the house. Askari was very much charmed with his gentleness. She treated him, in spite of her youth, as would a mother or an elder sister. As usual he talked as if the world, and he more particularly, were on the verge of calamity. She listened intently and comforted him with soothing words. I was glad, for Narinjan was very favourably impressed with her. And his face, generally so solemn, brightened up when he looked at her and talked — she was bewitchingly beautiful.

Often she would laugh and ask if she was not like a bird that had found her freedom. 'Yes, yes,' I would answer. 'And you fly so swiftly at times!' She looked at me, knowing what I meant, but turned it off with a ripple of merry laughter.

I wanted her to taste her freedom to the fullest. In the late evening I would drive her around the town before taking her home. One afternoon I took her from her house and drove thirty or forty miles, almost to Jullundhur. She was thrilled with the vastness of the Punjab and with the green song of the rural earth. All the way back in the car she sang spiritedly of the spring and the heaviness of the flowers and the fruit. The immense release of her personality through even the simplest new experiences was exhilarating.

I had not yet shown her the whole house. But one afternoon when she came early I took her to the roof terrace and to the tower rooms. She revelled in the vastness of the view, and thought the tower rooms romantic and toylike. Then quite casually I took her down to the third floor where three rooms housed my library. I remember assuming that she would just pass through the rooms scarcely noticing them. But I will never forget her overwhelming response to the rows and rows of books. She was amazed, awed, and ecstatic. She looked at me, her face alight and yet serious, and exclaimed, 'This is the most wonderful thing I have seen! And you, Gyan, you are the most fortunate, wonderful man to have all these books to read. What an immense universe of knowledge at your command! Have you read them all?' I laughed and had to confess that I had not. Then I led her to my study on the lower floor. Its atmosphere of comfortable studiousness thrilled her.

'How can you tear yourself away from here, Gyan? How can you let me distract you from all this wealth?' Her face now was almost solemn.

I smiled sadly. Yes, it was wealth, but how could she forget the warm pulse of her blood and the feeling in her being for tenderness and love?

That evening Askari kept tripping upstairs to the library and to the study. She could read Urdu and a little English. She busily deciphered what she could. She was much too distracted for love-making, but it was not till late that night that she could be persuaded to leave, exhausted but wildly enthusiastic.

The next day she sent word begging to be excused, but

she would come the following evening — she was exhausted and had to rest. I felt it was well that she should relax and recover her balance.

The following evening I expected to see an Askari much more like her old self. But she was more excited than I had ever known her, and she immediately rushed upstairs to the library. When Rustam was ready to sing she said, 'Let' him begin. You listen to him and I will come down when I can.' There was no stopping her.

Again for two or three days she did not visit me. After that, when she said she would dine with me, I no longer knew in what mood to expect her. She was calmer than she had been for the last week or so, but it was not the previous poise of a woman confident in her beauty and charm. Instead there was a grave look on her face. She smiled as if she were sorry for me. I glanced at her again, and in her eyes was the sort of look that goes out to a man who is doomed. I was startled, and asked her if all was well. She merely nodded and again smiled her strange smile.

I was surprised that she did not immediately go upstairs to the library. Her swift enthusiasm had exhausted itself, and now her mood was changing, I thought — wondering also whether having known me as a lover, and then as the householder, her interest in me was over. But after she had sipped a scented soft drink she asked me, in a manner befitting her grave look, to accompany her to the study.

She sat down in a large leather chair, a hard concentrated intensity shining in her eyes. It was a look I had never seen on her face. Then she told me what had come over her.

'Gyan, I have not been truthful with you these last few days when I said I was too tired to visit you. There have been other things that I have been doing with my freedom.' She stopped, and I felt a chill go through me from head to foot. Had Ranjit returned? Or was there someone else, younger and more vigorous than I?

She went on, setting that fear at rest. 'I invited to my house two teachers from the local women's college. I told them of my poor education, but said that I was fired with a desire to read, to learn to express my thoughts, to discourse with other people. I begged them to teach me. They realized I was serious. Yesterday and earlier today we went out and bought the books with which I will commence my studies.' She raised and clasped her hands, and her face struggled to suppress a too vivid uprush of joy, perhaps because she did not wish to hurt me.

Then she continued to tell me of her new plans. 'They tell me that if I work diligently, not losing time, in two years I can finish the high school courses. I intend to take all the necessary examinations and to go through college. And then I will really begin my life. Till then, I will be engrossed in the first steps.' She stopped and looked down. She must have realized that I was baffled by all this, and she hesitated to tell me more. But after a few moments she spoke again, more slowly than before.

'Gyan, I can never tell you how grateful I am to you. By bringing me to this beautiful old home of yours you have truly freed me. You did not know, nor did I, how this freedom would affect me. But it was right that I should be freed, and it was you who did this for me. I know you braved all the social consequences which might have followed. I know how unselfishly you acted. Now that the bird has found its wings it must fly where its spirit takes it. I know I must do something urgently to fill the vast deep chasm of ignorance within me. My freedom urges me to do this. I know you will understand and forgive me, Gyan.'

Askari got up and stood before me. She looked into

my eyes, trying to steady me. I could see that she was determined. Her body had already closed. Her face was firm, and her expression impersonal. We went down, and after dinner — which she ate in silence — she bade me farewell and left.

It had all happened so swiftly. I knew she would not return, but somehow that knowledge was immobilized by my longing for her and the wild hope that she might change her mind when she faced the loneliness and lovelessness of her decision. That whole night I sat in the drawing-room, my eyes fixed on the wide open front door. But only the cold morning air entered and drove me to my bed.

That was how the freed bird, Askari, who in her confinement had brought me the full sweetness of love, flew away from me. She was deeply sincere in her passion for education. Eventually she did get herself a college degree, and I am told that now, though she still enjoys music and poetry, she teaches in a private school for girls at Amritsar.

Part V

ASKARI'S sudden flight, just as our relationship was developing to a fullness of affection and confidence, left me like a tree stricken with lightning — reaching up starkly, a great twisted dry bone.

I remained indoors most of the time, hiding my feeling of hollow emptiness. I ate without desire for food, slept without feeling the need for rest, and in fact existed without any desire for life. Even in respect of Askari my feelings were numbed. I felt no reproach against her. I recollected her description of herself as a bird that naturally had to fly away once its wings had grown. It was as if her wings had been fashioned within me, and on their take off had torn out of me the vital spark of life itself.

For days this petrified state continued, till I began to accept it as the state of being I had reached. Then it seemed to thicken, to create a total opaqueness within me which seemed final and was without any relationship with my surroundings. I felt removed from everything around me; and recollecting also the failure of my relationship with Jennifer, with my wife Basanti — who had died, just as it seemed that we were coming together — with Pushpa and now with Askari, I felt I was confronted with enough data to establish the futility of my attempts to enter into satisfactory relationships. Perhaps this was merely a rationalization, for as I have said I already felt completely removed from my surroundings, and there was no question of my ever attempting to form new relationships.

If I was now a stricken, gnarled, bare tree, and was

already remote from all that was around me, what was I doing here in the family mansion at Amritsar? Besides, had I not experienced the workings of a law of failure in my relationships? And the house itself was as hollow as my own inner self. What was I to do? Of course, I could pay my sister a visit; or take a place at Lahore or Delhi for a change. But the thought of these possibilities simply intensified the feeling of deadness within me.

Narinjan still came and saw me almost each morning. He was again disturbed by thoughts of future mishaps. His was a strange kind of pessimism. He never complained of the past, or even the present, but he was obsessed with the terrible things which were about to occur. When they actually occurred he seemed to be able to absorb the impact. Perhaps that was why he still looked very much the same as thirty years ago. Strangely enough this was one of the rare occasions on which, though still complaining of the evil propensities of Jupiter, he predicted that a bright and glorious future awaited me personally. I smiled wanly. At no time in my life could the news have been less meaningful or possible of fulfilment.

During my talks with Narinjan I learned how much the war had stimulated the demand for timber. Prices had risen sharply and there was talk of their being controlled. Narinjan said he would welcome the step. In brief, I assured myself that his business affairs, and mine too — though the latter seemed now to be of no consequence — were in good condition.

I knew, too, that the paper mill was doing excellently, and my younger son was well settled in Delhi. Moreover, I had already transferred some of the property to his name. Thus I had carried out my responsibilities to those who could make claims against me.

Calmly I thought over the situation and took my decision. There was nothing more which I had to attend to in the sphere of worldly duties. Nothing held me to my present scheme of things. On the contrary I felt so stricken and alone that I thought it perhaps my duty to remove myself from it. Maybe our ancient sages were right: a man after fulfilling his worldly duties, when he was just past the meridian of his life, should don the habit of a hermit and retire for a period of meditation.

One evening I wrote brief letters to my son, to Narinjan, and to my sister, and told them not to be alarmed at my protracted absence. I asked Narinjan to take over the house, and I left him a power-of-attorney to cover all my own business interests. That evening I told my oldest servants not to be anxious on my account if I was absent for some time — Narinjan would look after them. I gave them my letters to post — I knew that they could not be delivered till some time next day: it was already too late for them to be cleared that night.

In the small hours of the morning I put on simple white clothes, and wrapping myself in a white shawl I set out to walk the fifteen miles or so to the grove of trees between Amritsar and my farms — where I had occasionally rested. I had liked the face of the tall kindly looking Sadhu who resided there. I felt certain he would advise me as to what course I should follow. I would get there shortly after daybreak, when he would have finished his morning prayers and would perhaps be willing to converse with me. None of my friends or relatives would suspect that I could be so near Amritsar.

When I reached the Sadhu's grove, tired and hungry, the sun had just risen, and its first level rays were making pleasantly warm golden channels under the trees. I stopped a moment, wondering whether I should wait till the Sadhu came upon me in the grove or whether I should go to him. The doubt was quickly resolved. I had come deliberately to this grove because he was here; it would be

182

a prevarication now to wait till he chanced upon me. I went to the great peepul tree, towering above all the others, under which he had usually sat. At its base was an earthen platform, about fifteen feet square and about eight feet high.

I climbed the earthen steps to the top of the platform. The Sadhu was seated on it before a dying fire of three slender logs. He sat very still. His slender handsome face was composed and pale. His grey-green eyes were open but seemed to have no sight for me. His heavy grey hair was piled in a knot on his head, and his grey beard was knotted neatly under his chin. Loosely thrown over his shoulders was a saffron-coloured shawl of handwoven cotton cloth, and he wore a small piece of similar saffron material about his loins.

Seeing that he was meditating, quietly I sat crosslegged on the platform. I knew I could not even greet him while he meditated, but the pleasure of seeing him was so great that an involuntary smile spread over my face. It was not till about five minutes later that I realized I was still smiling, and that, seated almost directly opposite the Sadhu, I might be disturbing him. Quickly I suppressed my smile.

After about an hour the Sadhu moved his hands under his cotton shawl. His body stirred and I knew his meditation was over. For a moment he closed his eyes, and then they opened, very bright and alive.

Looking at me, he said in a low full voice, 'You have not come on your horse today? And so early.'

'Sadhuji, I have left my horse, and all else, and have come here.'

'What else have you left?' he asked.

'I have left everything. It is not very much. Just a house, land, and other possessions. My children are grown and settled.'

He smiled and said, 'I hope, friend, the horse is well taken care of.'

His eyes responded to the answering twinkle in mine. Then he talked in a way which seemed to be as much for the trees and himself as for me. 'He has come here, but he has not fled. No one flies. Each moment, the world is destroyed and is reborn. Each moment is in balance. Comings and goings have their significance for us, but they alter nothing.'

Later I found the Sadhu would often talk as though his words were part of the shadow, the sunshine, and the fire. On this occasion he fixed his gaze on the smouldering logs in a way which suggested that perhaps it was the fire which told him what to say.

He asked me to put the logs together so that the hot ashes and the burning coals, which lay in the centre, could ignite them again. Then he inquired about me. His questions were brief, but I realized he felt he should know who it was that had come to share his grove. I must have spent two hours talking to him about myself, and then I was exhausted by the effort and by the strain of my journey, and by hunger. Seeing that I was weak, the Sadhu turned about and stooped to enter a small door behind him in a moundlike earthen structure at the southern end of the platform. I could hear him descend a flight of wooden steps. In a few moments he returned with some parched wheat and a small pitcher of water. He brought them to me and told me to eat.

While I ate I noticed again how still and upright the Sadhu sat. I thought he was meditating again, but later I learned that he never wasted a single movement.

When he saw that I was refreshed he arose. 'Come,' he said. 'I will show you your abode.'

He was tall and spare, and his long handsome face was finely made. His eyes were bright, almost like an animal's or even a bird's. Often I thought he must see what the birds saw and which passed by me unnoticed. He took me to a whitewashed room built on a low brick platform under one of the trees. Its walls and roof were of clean earth. It had one door and a window, and no furniture of any kind. He left me to rest.

Late in the afternoon I awoke, my body stiff from having lain on the brick floor. I sat in front of my door very still, already learning from the Sadhu, and gazed through the trees. They were motionless and heavy. Occasionally a bird chirped, but otherwise it was an hour of silence. I was at peace, and the little whitewashed room behind me was, I felt, an ample retreat from the outer world. I was ashamed that already hunger was gnawing at my stomach again.

The tall figure of the Sadhu, moving very gracefully as if engaged in some quiet ceremony, came into view. He walked with an economy of movement which gave him a striking carriage, and when he was seated he had great poise. He came towards me and said, 'Come, I will introduce you to certain places which you will need to know.'

I followed him in silence, first to the tank. One side was formed by the slope of a hillock as was part of another side, a low brick and masonry wall completing it and a third side. The remaining eastern side consisted of brick steps leading into the water. The Sadhu led me down the steps. He removed his wooden sandals and went to the water's edge. At the end of the steps he lightly brushed the surface of the water with the side of his hand. I saw the water there tremble faintly, like the beginnings of a simmer in a pan of water on the fire.

'At this place there is a gentle spring that freshens the water. One can wash here and feel cleansed.'

Then he took me round the tank to the other side of

the hillock. Under a jutting rock he showed me a small trowel which he kept there. 'With this,' he said, 'I dig a small hole when I need to ease myself.'

He led me into the grove again at its southern end, thick with mango trees. Each year the owner sold the fruit to a contractor, but he permitted the Sadhu to take a handful of mangoes each day. 'But I seldom exercise this privilege. I did not like the way the man bestowed it on me. I did not ask for it, but he sounded as if I had petitioned him.' Though I delighted in eating ripe mangoes, in all the time I spent with the Sadhu I did not take any of the fruit.

As we went through the two acres or so of trees I was still hungry, and I hoped he would tell me how we procured our food and drinking water, but he mentioned nothing about the matter. It was time for his meditation. He said I could come and sit on his platform at any time of the day or night, provided I did not expect him to interrupt his meditations.

I parted from him at the foot of his platform and returned to my own lower structure. There I sat down and wondered whether I would succeed in attaining the stage of effortless movement and frequent spells of meditation. I tried to sit still and let my mind dwell on the peaceful influence of the oncoming evening. But I would find myself moving a limb or glancing at a bird returning to the grove, and soon my body began to sag. I smiled to myself, and excused my failure on the ground of intense hunger. My body was warm with the faint protest of my undernourished bloodstream, and my head felt as if it were soaring away from me. Besides, who could help being diverted from meditation by the intemperate calling of the birds in the mango grove? Their thunderous contention was accompanied by spirited jostling for the most leafy shelters. Then, all of a sudden, there was silence except for an occasional affectionate twitter in the midst of a feathery embrace.

Before the distilled violet light of the evening flowed entirely away I climbed the steps to the big platform and sat in front of the Sadhu. A young turbaned peasant was seated on the platform. He was lighting the evening fire, and by his side, within reach of the Sadhu as he sat upright before his diminutive doorway, lay a flat wicker tray covered with a village kerchief in a red print design. The Sadhu reached out to it, and uncovering the tray he passed it to me, nodding hospitably. On it were laid raw carrots, two thick chappatis, two bowls of curd, green leaves, and some soaked pulses. I took a leaf from the tray and nimbly arranged on it some of this dazzling array of comestibles. I passed back the tray and looked at the Sadhu. He nodded to me again, and I ate a sweet and satisfying dinner. The Sadhu also ate, and what remained on the tray he placed on a fresh green leaf which he put by his side. There was one glass of milk, which he handed to me. Obediently, but also in response to my still clamorous appetite, I drank it down. The young man left a pitcher of water, and quietly went down the platform steps.

The Sadhu rose effortlessly, and slipping on his wooden sandals, he walked to the water's edge. I heard him splashing his face and rinsing his mouth. A little later he returned. I felt already that this was part of the way of life of the grove, so I rose and followed his example.

When I returned he said, 'He is not always so quiet, the young man. He likes to come and talk. The village headman tries to give others the duty, but he insists on coming every day. Two days he had to stay behind because the police had to question him about a crime — his regular visits here proved to be a help; they were an effective alibi. He asks me the most unexpected questions.

One day, he said that as all things have their meaning then surely the leaves chattering in the breeze must say something. He asked me to tell him what the leaves in this peepul tree above me were saying that evening. Truthfully I said to him I knew what they were saying, and I also knew what the other trees were saying. "Then tell me, Sadhuji," he begged. "Listen to them, and you will know," I replied. But he laughed and thought I was playing the fool with him. He could not understand that I knew what the leaves were saying to me, but it was not necessarily what they were saying to him.

'At other times he tries to involve me in village disputes by telling me of supposedly hypothetical cases and asking which side is in the right. Again my answers baffle the poor boy. Generally I find that both parties are very much to blame. But still the boy comes; perhaps just because he finds me baffling.'

I wondered whether I dared ask the Sadhu whether he would instruct me. Perhaps he would reply that I should find my own way. I knew, too, that that was what I must do, but at this stage I did not feel equal to the task. So I shirked the question; and besides, he did not ask me when I planned to leave. If he assumed that I was staying, and the present terms were acceptable to him, it would be foolhardy on my part rashly and hurriedly to alter them.

That first week in the grove I was mainly an observer of the dense green trees lifted on their dark sensuous trunks, the shimmering water of the tank, and a single line of saffron formed by the Sadhu's movements criss-crossed against the trees, the water, and the earth. Once a day we met to eat, but he seemed to remain in the remote regions of his trance and meditation and never uttered a word after the first evening. As the week closed I did not know whether this introductory state would continue or whether another phase would commence.

On the eight day, while the Sadhu was passing my hut — as he often did — he stopped and called to me, 'Come with me, friend.'

We went together and sat on his platform. His eyes, which for seven days had been lost in a mist, were piercingly clear. His cotton shawl fell loosely from his shoulders, displaying his bare chest and arms on which the taut smooth skin covered his wiry form. He was intensely alive as he sat before me. I looked with admiration at this man of, I supposed, sixty years or so. He was smiling at me, and I knew he had a purpose in asking me to come with him. But before he could express it I could not restrain myself, and exclaimed rather than asked, 'Sadhuji, why do you live?'

It was an impertinent question, and as soon as I had asked it I wished I had not. Surely, catching him off guard, it would annoy him by thwarting the flow of what he had intended to say to me. His keen eyes twinkled for a moment, but he smiled again and replied:

'Of what special significance is my life or your question? I am part of the pleasure of the Creator.' Then his smile widened, and he added banteringly, 'Do you think I fail to give Him pleasure?'

I was abashed, but pleasantly, and replied in his own vein, 'I am not the Creator and do not know, but if I were I would accept your explanation of existence! But forgive me, Sadhuji. It was a foolish question.'

'No. No, Narayana. It was not a foolish question. Questions seldom are, but answers only too often. Now let me ask you a question. You have had enough time to observe; do you intend now to stay here for a time?'

I nodded, and wanting to push further than even his question, I said, 'And to seek instruction from you.'

'Whatever I can give you will be yours. But you know the law of this kind of life? I cannot say that I will remain

189

in this grove even for a week. At any moment I might feel that I had to move elsewhere; or I might be here for many years. And you, too, must feel the same freedom, Narayana.'

I smiled: immediately after I had asked him why he lived he was calling me 'Narayana' — the Immortal One.

It was agreed that he would instruct me commencing from the beginning, for, excepting the stories of the epics and some snatches of the Bhagavadagita — which I had learned from my mother — I was ignorant of all the sacred books and philosophies.

Then in a gentle voice and in a tone of suggestion he added, 'Two things I would bid you do. One, simplify your own life, and then simplify that of others by setting their minds at rest.'

'Both suggestions, as you put them, seem most desirable,' I replied, wondering what precisely he had in mind. But it was soon apparent. The first was that I should don a saffron-coloured shawl and loincloth, such as he wore. They would give me anonymity — which I desired — and would remind me, by their cheerful glowing colour, of the joy of creation. He had been given a new set by the villagers. He went to his cell and brought it to me. Every day, or every other day, I would wash them, and soon someone would give me a spare cloth. He told me to wash my white shirt, shawl, and trousers and leave them with him. Later, I saw him give them to a poor traveller who passed by the grove in tatters.

His other request meant that I was to write to my closest relative or friend and tell him that I had adopted *sanyasa-shrama* — the way of the hermit — and was faring well. They should not be left in uncertainty as to what had become of me. It was possible that, in time, it might be right for me to reveal myself to them; but even if it should not they must at least be made aware of the decision I had taken. He asked me to date my letter and to say that its postmark must not lead them to think I was in its vicinity. He would undertake to give the letter to a person who could be relied upon to post it at Delhi or Allahabad. My caution about the postmark would deter my people from setting out to search for me.

The next morning I was ready, in my saffron loincloth and shawl, to cross the threshold of the new venture which had taken me away from my home in Amritsar. That morning I felt I was truly stepping out from the big gateway of the family mansion. Its old rust red bricks seemed to flush, making me feel acutely self-conscious and diffident. But at last I felt I had moved from the house to the grove.

Soon I realized why the Sadhu had smiled when I had asked him to begin his instruction with the basic religious texts and then to lead me on to the philosophies. Gradually it became clear to me that learning the books, and even the philosophies, was the simplest part of the instruction which I sought of him.

Each morning I awoke when the single sentinel birds cut through the paling light with their tentative calls, so different from the sensuous evening clamour of their homing flights. Like little silver knives lit by the first rays of the sun, these early bird-calls pierced my ears and slit open my eyelids. When I arose and reached the tank the water was beginning its feline trembling at the first touch of the sun's light. After my ablutions I returned to my hut and sat on my low brick platform.

Then the birds would begin to rise from the trees to set out on the day's forays, and I would hear the resonant voice of the Sadhu intoning. It was the signal for me to go to his platform. I would sit before him, across from the fire, silent while he intoned from the Upanishads, and my ears opened to the chiselled intricacies of sanskritic sound:

'Sthane rhishikesh tav prakirtya
Jag prahrishyatyanurajyate cha.'

After he had intoned a whole Upanishad the Sadhu
would look at me, his eyes two pinpoints of sharp radiance,
and eagerly he would exclaim, 'See, Narayana, how the
whole Upanishad has its own unity of sound!' Then he
would intone the whole text again so that my ears might
be filled with the rhythmic precise pattern of sound.

Then he waited, and his face became serious. In a
deeper, slower voice he repeated slowly the first verse,
nodding to me to follow after him. And so I learned,
verse by verse, in between listening to his commentary. The
explanations he delivered poured from his lips in a stream
of beautiful words as if there were two equally clear springs
in his mind: one Sanskrit-filled, and the other, Hindi.

'Distrust those who say to you, "this, I tell you, is the
meaning". For me, each day the meaning changes, and
for you the word will have its own special evocations and
significances.'

We had no set period for the day's instructions. He had
his own way of sensing when fatigue was approaching
either or both of us, and then he would stop. One day he
said, 'Let us stop, Narayana. Your eyelids are beginning
to quiver.' Then I thought, so I know now how he decides
when to stop; but another day he would say, 'Your ears
are pink', or 'your toe twitches'. So I knew only how
intently he watched me.

On the winter mornings we commenced when the hoar-
frost was rising as thick as heavy smoke so that our grove
became like a green dell curtained off from the rest of
the world, and the Sadhu's voice almost echoed from the
rim of our own cold drier dome of air. On those mornings
the earth was like frozen steel and quickly drained away
my warmth, leaving me numb to the bone. The Sadhu

noticed that something was the matter, and soon two little mats came from the neighbouring village for us to sit upon.

In the evenings after we had eaten and returned from washing at the tank we sat before his fire, and he would talk to me under the stars, telling me of his years in the mountains, of his own guru, or of the joy of living the way he did — unencumbered, free to go where he wished. But some evenings I would know he did not wish me to come to his platform. He would sit still in meditation. I felt he frequently sat thus through the whole night. On the following mornings his voice would be softer, and after midday he would rest, sometimes sleeping as he sat upright.

At first my unpractised memory fared very poorly and I learned slowly. But I persevered, repeating the verses through much of the day. Eventually they became part of me, so that I could concentrate on a particular Upanishad till its stanzas came into the focus of my consciousness and poured from my lips.

But there were other things to learn. Through the winter I had to sleep on a thin pallet of straw on the bare brick floor, covered only with my cotton shawl. The Sadhu encouraged me.

'Move gently in the day, preserve your energy, breathe regularly and slowly, and then at night your body will radiate a layer of warmth around you.'

He laughed at the grimaces I made when first I tried to eat uncooked food. To help me he soaked the wheat, other grains, and lentils, till they sprouted. Then they were soft and sweet. But cooking he would not hear of. 'What wisdom is there in letting the flames consume half the strength of our food?'

Once a day we ate, and sometimes in the mornings we were brought a glass of milk each. My body wasted away, and I wondered how long the process could go on. My skin hung loose about me and my bones protruded in ugly knobs. But after a few months the loose skin started to become taut and the muscles beneath it rippled when I moved. The Sadhu looked at me approvingly and said, 'Narayana, now at last you can become acquainted with your muscles. Soon you will learn to get them to do your bidding.'

He beckoned to me and asked me to hold his forearm. Without moving his arm or his hand he flexed and un-flexed the muscles under my hand as he pleased. Then he put my hand over his heart and I felt the steady beat. He sat very still, and I could no longer feel anything. I looked at him with concern. About two minutes later he smiled, and again I felt the soft thudding of his heart. He shook his head. 'Do not ask me how it is done. Not yet. Do not, Narayana, ever covet these powers. No power is worth possessing. But, if one has need of certain faculties, one is obliged to develop them.'

It was not the learning of the texts, nor the eating and sleeping, which were the most difficult parts of my instruc-tion. How was I really to lift myself into a new zone of experience by meditation? For hours I would sit still in front of my hut — but nothing would happen. During the day as I gazed upon the trees I would often become drowsy, and the only way to keep awake was to go to the tank and splash my face with cold water. That would alert my being to the smell of the mango blossoms, or to the sight of the gay, bright yellow mustard fields; and I would stroll around the grove, happy in a simple idyllic way, forgetting all about meditation and concentration. Some-times as we ate in the evening the Sadhu would ask, 'Narayana, has the day been pleasing to you?' I would confess to him how the joy of meditation had eluded me, and how other simple things had pleased me.

'That is good, Narayana. Never forget that the main

194

objective is the pleasure that brings no pain. Do not be too intent on this way or that to attain it.'

So I forgot about meditation and went on with a life mainly of study and observation. It took us about two months to cover each of the principal Upanishads. When we came to the very short Isa (Oh Lord! Lord!) and Mandukya (The Sacred Word) Upanishads, I thought we would complete each of them in a few days. But every word seemed to soar away from us or fall deep into the depths of the ocean. Our journey through these brief Upanishads was the most strenuous of all.

During my first two years in the grove I would retire to my own hut when visitors came to see him. But at the end of the second year we had almost completed the study of the thirteen principal Upanishads, and he would call to me when others were present, 'Narayana Sadhu, will you consent to aid us with your views!'

Thereafter I was present at many of his talks. People came from the surrounding villages with their problems. Some just wished to sit in his presence; others asked advice; others tried to convince him of the righteousness of their causes and to ask the assistance of the supernatural in convincing a magistrate; women asked for the birth of sons; some came and asked that their excessive passions be quietened. The Sadhu would deal with all his callers quietly and with humour, so that sometimes there were bursts of laughter. But never did he explode with temper or annoyance, even when a trying caller was taxing his patience. Often he would say to me, 'It is those who come and sit silently who perhaps gain something. The silence helps them, and perhaps I do in some way make manifest the delight of Brahma in his creation. But the ones who talk get nothing. What do I know that they do not? Indeed, about the subjects they raise, they naturally know more than I do. They come to convince me; but I am never convinced wholly of anything, and that hurts them, poor folk!'

It was not only his enduring gentleness, humour, and modesty that I witnessed. Most impressive was his amazing dignity. He sat like a king — straight — and his eyes radiant. I had thought he was ten years older than I, but he told me he was a good twenty years my senior, and yet each day he seemed straighter and more radiant. I suppose, in my admiration for him, I was somewhat carried away, but he was undoubtedly a most striking man.

In the third year of my stay at the grove the surrounding villages became agitated. The visitors now sought our aid and advice not for their personal lives but inquired whether they would be swallowed up in war and strife. The Sadhu shook his head. 'Strife? I do not know, but I hear men are creating strife. Nations are at war. The Government of men from afar is tardy about letting the people conduct their own affairs and the people are restive. Hatred is being stirred up by self-seeking persons and by those who use religion to divide and not to humanize.'

The people had risen against the Government which acted ruthlessly with one hand and distributed largesse with the other — giving its friends and the sycophants jobs, contracts, titles, and grants. The people were confused by the fact that scarcity of goods, misery, suppression, killing, and jailing went side by side with plenty, the building of fine houses, and the mushroom-like growth of the new rich. And now there was talk, too, of religious strife.

Late that summer of 1942 there were night meetings, under the mango trees, of groups of agitated men. The first night they came up to the Sadhu and myself — we were both seated on his platform — and politely asked whether they might hold their meetings under the trees.

I wondered whether the Sadhu would raise some equally politely phrased objection, but his approach was different.

'Certainly,' he said. 'If your friends are holding meetings which you think will do good, you should hold them wherever you think is best.'

'Thank you, Sadhuji,' replied their leader, joining his hands respectfully. 'These meetings are for the good of the country. We will all benefit from them. A new age is before us, Sadhus!'

When they left the Sadhu said, 'People so often make the mistake of thinking that we are recluses who want to avoid the coming and going of people. That is not so. We live in these quiet places so as not to be in the way of other people. For a year I lived in a room in Delhi. But the city was overcrowded. There were many workmen in the city who had no place to stay. It was clear to me that if I moved out at least one person — or perhaps a man and his wife — would have a room in which to live. I was not working in Delhi and there was no particular reason to remain there, so I left.'

I realized that I had fallen into a kind of bondage of seclusion from which the Sadhu's few words released me. I felt an inner potential of movement stirring within me, and the scent of freedom was in my nostrils.

Towards the end of my third year in the grove a development occurred in my attempts at meditation. First I found that I could sit still without quickly developing restlessness — it was as if at last I was ready and expectant. Then even the stillness became remote from me. I felt myself suspended in space; then the earth was no longer different in substance from the atmosphere around it and I was suspended within it; I was skimming the waters of the Ganga, and rising above the clouds. Thus it was that I began to experience the meaning of many verses in the Upanishads:

It moves and is motionless
It is far and remains at hand
It is within everything
And it is outside all of this....

Sitting, he proceeds into the distance.
Lying at rest, he goes everywhere....

I do not know for how long I sat rapt in meditation that
first time. The Sadhu looked at me strangely when I went
to him after emerging from my trancelike state. He knew
what had happened. 'Narayana, you know now the bliss
of life; the free awesome contemplation of, and pleasure in,
the creation. It is even more staggering than you have so
far seen! You do not need me any longer, Narayana.'

I smiled at him. I was breathless — not just in a physical
sense, but because of the feeling of remoteness. When I
regained some control of myself I said to him, 'Sadhuji,
this is the beginning. Let me stay with you yet some time.
I see in your eyes that you soar a million miles away. I
want to be able to come closer to you before I take my
leave of you.'

At the end of the year I completed my course in the
Ramayana, learning all its 24,000 verses. And in that
year my flights in meditation ranged farther and farther
into the dark distances of space. As I felt myself soaring
upward the blue of the sky deepened in intensity till it
became as black as night. Groping my way in it I would
come upon an awe-inspiring breathless void, and then,
suddenly, I would feel myself descending into the earth.
Then perhaps I would be drawn through the trunk of a
mango tree, up its tender branches and through its green
leaves, to burst again into space in the scent of its blossoms.

Meanwhile I became aware that the glow on the
Sadhu's body had acquired a greater radiancy. In certain

lights it was almost transparent. His frame remained as I had always known it — lithe, and wonderfully poised. But in the second or third month of my fourth year with him he said to me, 'Narayana Sadhu, my mind and body have developed a lightness. They will not be heavy to carry. I think this is a sign that I must move on from this grove. I do not know where I will go, but my eyes turn at present towards the line that divides the south from the east.'

I bowed my head. The news affected me deeply, and yet it sounded familiar to my ears, as if it had already reached me some time ago.

When I looked up at him he said, 'Undoubtedly you knew that this must happen. It is good to know and also to have reticence. Knowledge is like the light that glimmers on that tank — the light has a thousand facets, and yet releases a thousand facets of darkness. One sees all of them, but they baffle one's mind and one says "it is a tank of water". So, even though we will separate now and say we have parted, in the inner recesses of our knowledge of life we will know something about each other. Only, remember one thing, Narayana: the pleasure of the Lord in his creation.'

That night we ate joyfully. He was in a gay mood and remarked with pleasant laughter that I·still grimaced involuntarily while eating some of our uncooked diet. 'But if you ever travel and have to search for roots and leaves you will become truly acquainted with the art of nourishment, Narayana.'

After our meal we washed at the tank and returned to his platform. Soon we were both lost in meditation and I felt myself soaring farther than ever before. ·

The sun's rays were striking my face when I emerged from my trance. I was alone on the platform. I never saw him again.

★

For a few months I remained at the grove, going over the Upanishads and the Ramayana that I had learned from the Sadhu. Every morning I intoned from them, and in the evenings I meditated. I was pleased that I was able to recite all the Upanishads and the Ramayana. Often I heard the Sadhu's voice intoning them with me. And though my meditation was, in a sense, utterly solitary, I sometimes felt him brush past me on his own lone ranging. For a moment my course in space would be lit by his radiant eyes turned on me, and I would see his face break into a smile of great happiness.

It lacked but a few months to four full years at the grove when I completed the recitation of all I had learned. I felt that my term of residence there had been completed. I did not spend much time in thinking about the direction or objective of movement. I knew that when the time came I would go, and that was all.

One night it rained heavily but briefly. It was a sharp, late monsoon shower at the very end of the summer. The next morning the matutinal orchestra of the birds was a little more vigorous than usual. As I intoned my morning recitation my nose tingled with the fresh scent of the air which came to me strongly with the rays of the rising sun. I completed my recitation and sat very still, allowing myself to feel only the strong fresh scent. It seemed to draw me to the east. I arose and struck out eastward from the grove.

The rhythm of steady walking was so strange that after half an hour or so I felt giddy with it and had to sit down to rest. My muscles were stiff when I got up, and thereafter I walked in short spans, choosing the narrow cool paths between the fresh green fields. The farmers raised no objection when I took vegetables, corn, or ears of rice as I went along. I rested for the night under any large tree that happened to be near by at sunset.

During the first few nights I was too fatigued to meditate and was troubled by my lapse; but as I grew accustomed to the rhythm of walking my fatigue disappeared and my periods of meditation became an exciting plunge into a different world.

As if to compensate for the movement of my journey the world I now entered in my meditation was one of linear stillness in which the passage of time was accelerated, so that I would see the rise and fall of a city in a space of half an hour, or a flower would burst open from its bud, extend its petals until the delicate skin stretched to the breaking point and withered. A moment later it was dead and its seeds fell to earth. I was still, space was still, but time sped past me.

Each day, after an evening or even a whole night of meditation, as I walked I felt acutely the meaning of motion in contrast to my inner life of stillness. The two together brought a new understanding of the creation, which I acknowledged to myself with a smile. I had lived for about fifty years without experiencing either stillness or movement with the vivid intensity which pertains to them. The Sadhu's words came to me. 'Narayana, remember only the joy of the Lord in his creation.'

There were days when I did not walk, but remained under a tree, reciting and meditating. It was about two months after I left the grove, for the autumn crop had been harvested, that I reached the village of Parmanand — the Abode of Great Bliss — near the foothills of the Himalayas. I sat under a great fig tree to the east of the village, from where I could face the mountains. In this way I could reacquaint myself with the mountain country and prepare to travel into it.

But my quiet personal preoccupations were disturbed the day following my arrival at Parmanand. The villagers came to beg my help. An epidemic of high fever was raging in the village, and already it had killed many of the men and children — for some reason the women were not attacked. I told them that I was not endowed with powers which could help them against the disease. They looked at me, downcast and troubled. Then one of them said, 'But there is something in his face which gives us strength.' This announcement came as a surprise. I had heard no comment on my face for years. I hardly knew any more how it looked. Sometimes I saw it reflected in the water — my nose and eyes much more pronounced than I had thought them to be; my face covered by a greying beard which, like the Sadhu's, I had knotted under my chin; and my head a mass of grey hair which, again, I wore in a large knot.

I was about to protest that there was no special power in my face, but yielded in silence to the chorus which greeted the announcement. 'Yes! It is so! It is the face of Vishnu, who has come to protect us!'

After some consultation among themselves one of the elder men said, 'Sadhuji, we are an afflicted people. Please consent to stay with us for a little while. That is all we ask.'

I could not decide whether I should impede my progress towards the mountains, and sat silently for a few moments. Then I said, 'Gentle friends, I do not know yet what I shall do. But at present I am here and I will not plan to leave you of my own accord. But you know, I must do as I am to do.'

They nodded, content with my words, letting me know, too, that they understood I could make no commitment. They brought me food, but at my request replaced it by the raw vegetables and soaked grain to which I was accustomed. Through most of the day the people came and went, one at a time, sitting before me while I meditated and recited. After three days with them the fever seized me. My joints flared up with pain so intense that I felt they were on the point of bursting. But I persevered with my meditation. And now, quickened it seemed by the blaze of heat that consumed my body, again I soared through space. The pace was faster than ever before. I sped beyond the darkness of outer space till I came upon a kind of granular liquidity, as if the evenness of the ether had disintegrated. At times I felt that this might be an indication that my own bodily existence was about to cease.

For three days the pain and bodily heat increased. On the third night my meditation sped me across the universe till I seemed to come to a bank of dark clouds. Into it I plunged and all seemed lost — a cold clamminess was suffocating me. But before it could stifle my breath I burst through the clouds, and instead of the granular blackness of the beyond of space a white haze of light was turning into a pellucid clarity, which seemed to intoxicate me. After that I remembered nothing till late the next morning. I must have slept soundly for many hours. When I awoke the fever was subsiding, and in the next two days it ebbed out of my system.

The fever left me so weak that my blood seemed unable to reach my numbed limbs. I ate very little, drinking only the warm milk that was brought to me. Three days later they brought me some heavy sweetmeats made in gay colours. I smiled, telling them I had not eaten such things for over four years. They did not press me, and I asked, 'Is this a day of festival?' I wondered at the cause of this feasting at a time of epidemic.

'It is a great day!' they said. 'Since you came to Parmanand there have been no deaths in the village. Many men and children were ill when you arrived, and others took ill at the time the fever struck you, but today none of them has any temperature.'

203

It would have done no good to repeat my earlier protestations. In view of what had occurred it would only have added the virtue of modesty to the power which they believed I possessed. Two days later the whole village came in a deputation — men, women, and children, including three blind and two lame persons and a baby barely twelve hours old. The people begged me to lengthen my stay with them. Again I explained that my intention was only to act as I felt urged — and that might lead even to my staying at Parmanand longer than they would want to have me. The mention of this possibility was greeted with loud protestations and shrieks of laughter. They ridiculed even the suggestion that they could even wish me to leave, and the argument as to how long I should stay concluded on this emphatic note.

In spite of the excellent health I had acquired living a life of physical hibernation at the grove, I recovered slowly from the effects of my illness. Even in meditation, now I would frequently find myself travelling through space in a supine position with the stars streaming over my face, or the earth rushing past — in the opposite direction like a solid massive stream. Thus the minor circumstance of a variation of my habitual posture revealed to me in a new way the amazing processes of creation, preservation, and destruction which exist side by side in our universe.

In the evening, seated before my fire, I would recite from the Ramayana, the Bhagavadagita, or the Upanishads. At first, perhaps influenced by the coincidence of the cessation of the illness with my successful personal struggle against it, large numbers of villagers would cluster round my fire to listen. But after two or three weeks the crowd began to thin out, and in another week there were many evenings when I was left alone. I confessed to myself, with some shame, that I was happiest on these solitary evenings. Through my remaining months at Parmanand the gatherings under my tree fluctuated — swelling when there was need for rain, or fodder, or more milk from the cows, and dwindling when all seemed to be going tolerably well.

When the summer heat began to raise from the earth the heavy decaying odour of dampness that had lain undisturbed under the earth's surface through the colder months, I left Parmanand. The scene of my departure contrasted strikingly with that of my arrival in the village. My leaving was witnessed by two village dogs, several boys playing *guli-danda,* and a woman returning from the well with three pots of water balanced gracefully on her head.

I walked eagerly towards the mountains, avoiding the village habitations. I slept under the trees, and ate roots and tubers and some grain and vegetables from fields along my way. When I reached the foothills, a tremor of excitement overcame me. It was in the early morning gloaming, before sunrise. The hills lay beside each other in intricate convolutions which broke the rim of the sky into a maze of beckoning pathways. And my excitement became almost a physical palpitation of expectancy when, looking at the mountain formations and their relationship with space, I felt I saw before me the very terrain I had sped over in certain phases of my meditations. Perhaps it is this uncovenanted recognition given to the initiated which has urged hermits and sages to go always a little farther in their search for a Shangri-la in the measureless mountain regions of Northern India.

The sight of the hills, so close to me, seemed to unveil the meaning of the kind of life which I had been living for the past five years. Now I realized what the Sadhu had meant when he had said that there was no fleeing from things. I had left Amritsar. That was true; but now I realized that the forces which had compelled me to leave amounted to an urge to live more fully and not to retreat from life. The life of meditation which I had sought, and eventually entered, was essentially the opposite of what I thought it to be. The man of affairs imagines meditation to be a state of idyllic retirement and a cessation of the adventure of life. I had discovered that in reality it was a life which demanded immense courage and a deep thirst for inner freedom — almost abandon. Meditation, apparently so calm a pursuit, had led me into an adventure in space which had taxed me almost to the point of physical disintegration. I remembered how I had often seen the Sadhu with beads of perspiration pouring from his brow and chest as he sat, his face seemingly calm, deep in meditation. This experience, into which I too had entered, was veritably an inner orgy, laying bare, in awful moments of vision, the whole magnificence of the creation.

The hills stood against the sky as the sculptured reflection of the adventure of meditation. They were to me as must have been to the first sculptor the discovery that he could give tangible form to the inner vision of life.

Because of my feeling of intimate relationship with the hills I moved in them slowly, exploring every yard. I went from tree to tree and covered but little ground in the course of a day. It was months before I had passed through the foothills and entered the vast Kangra valley stretching down from the first range of snow-covered Himalayas. Standing on the green skirt of those great mountains I felt already that I was being swept up into them.

The months passed in my slow peregrination, and it was not till the winter that I crossed the first mountain pass and descended to the beautiful town of Mandi, on the banks of the river Beas. There were beautiful houses in that ancient town, but its lofty temples sang of the

adventurous spirit of the people and at once endeared
them to me. There was a friendliness in the streets, so
that as I jostled among the crowds I scarcely remembered
the isolating ambience of my meagre saffron garb. The
timbre of life was akin to that which had been developing
in me the past few years. Here the people still expressed
much of the sweetness and pleasure of creation. As the
winter came to an end there was dancing and singing in
the streets and on the surrounding hillsides.

When, in the following summer, I left Mandi to make
my way upstream along the Beas, it was not without an
effort and a twinge of sorrow. I was shocked at this uprush
of emotion. As a hermit I was to practise perfect non-
attachment, and I had left the grove without any tender
lesions being created. Even when the Sadhu, in whose
footsteps I followed, had left there had been no faltering
in my strict adherence to non-attachment.

Telling myself that some passing weakness had overcome me, I soon re-established my tranquillity as I made
my way up the deep gorge of the river. The hillsides were
turning a burning copper under the autumn sun, and the
most insignificant shrubs and gnarled trees were aflame
with colour. Only the wild olives maintained their
placidity of silvery green.

It was in the following winter that a new experience
filled the hours of my meditation. I was inhabiting a slate-
roofed room near the confluence of the Beas and two
mountain tributaries: one from the north — jade green
and translucent, the other from the east — crystal clear
and reflecting now the gloom of the winter. The Beas
itself, at that point, was a proud recipient 'of tribute,
flecked with changing blues, greens, and browns in its
deep and spacious volume. As I meditated at this confluence — which, like all others, was sacred — I would
seem to be skimming the surface of three great converging

streams whose waters glowed in gold, deep orange, violet, rust, copper, and silvery blue. I was swept to the sources of the waters, where they emerged under glaciers of crystalline blue, copper, red, and golden ice. Such vividness of colour had never before dominated the world of my meditation. I welcomed this as a new revelation of the beauty of the creation, but expected it soon to be superseded by the black depths of space and the dense darkness of the bowels of the earth. However, as the winter progressed the colours increased in intensity, and the glowing waters though emerging from massive glaciers were balmy against the rarefied body that accompanied me on these swift journeys of meditation.

Though the beads of sweat stood out on my chest and ran down to my armpits and loins, I emerged from my trances exhilarated by the feasts of colour which they provided. Even when I intoned the epics and the Upanishads — as I did for several hours each day — it required intense effort to sweep aside the blaze of colour filling my mind.

In the spring I left my place of winter meditation, at the confluence, and entered the last valley of the Beas — the beautiful region of Kulu. The valley was wider now, and its cool pine- and cedar-scented air blew strong and fresh upon the fibres of my body. My meagre cotton shawl was powerless against the piercing winds, and each evening I sought a formation of sheltering rocks.

One evening as I came upon such a place I heard the sound of sobbing. Following it to a cluster of sharply protruding granite rocks I came upon a shepherd lad of about twelve, doubled up and weeping. I went up to the boy and saw that one of his legs was badly injured: the kneecap was swollen and bruised, and there was a deep gash on his thigh which appeared to be festering. Sensing my presence the lad turned his head, and out of a round, snub-nosed face, he looked up at me with his black slightly slit eyes. For a moment his look was one of antipathy, almost of revulsion, but immediately it gave way to an abject appeal which he reinforced by a louder sob and an attempted movement of his body, indicating that he could not raise himself.

I crushed several handfuls of astringent leaves into a spongy mass. Making a cup of a few large leaves I filled it at a near-by spring and washed the lad's wounds with my sponge of crushed leaves. Seeing that the astringent leaves drew away some of the clotted pus I made a compress with more of the leaves and applied it to his thigh and knee. I brought him fresh water to drink, and that evening he shared with me the nuts and tubers that I found in the vicinity. Immediately afterwards he fell asleep.

The next morning his wounds looked healthier, but he was still in pain and kept trying to reach with his hand the region between his shoulder-blades. I lifted his rough woollen coat and saw that his back, too, was blue and swollen, although there were no external wounds. That day the boy could only groan, eat, and sleep.

It was not till several days later, when his leg wounds were healing and he could sit up, that I was able to get him to tell me a little about himself. His name was Rumsu. He was an orphan and had come up from Kangra with a group of shepherds who were leading a large flock of sheep to the high pastures at the head of the Kulu valley and beyond. One evening, a few days before I found him, he had been rounding up some of the sheep and had slipped and fallen from the rocks above the spot where I had come upon him. He had heard the other shepherds calling to him in the distance, and though he had replied weakly the sound of the streams and the roar of the river had drowned his voice. The men must have looked for him but apparently gave him up for lost, thinking perhaps he had been swept away by the swift river.

After a few days, having responded to my treatment, and refreshed by such food as I could find for him, Rumsu became cheerful and smiling. He knew that he was recovering, and his eyes shed the glaze of dejection that had covered them. He seemed to enjoy any intoning of the sacred books and responded by singing snatches of songs at moments when he knew I was within earshot. But when I asked him to sing to me he shook his head and smiled shyly.

About two weeks after I found him, Rumsu was able to limp about the neighbourhood, but feared going anywhere that involved stretching his limbs, so that it was too early to consider what I should do with him. And just when I thought he was well enough for me to talk to him about his future, for no apparent reason he took ill with a temperature and an ague. His body exuded the damp, lukewarm odour of malaria. The illness must have erupted because of the debilitation which had resulted from several months of undernourishment. My scant diet was obviously inadequate for his youthful needs. His temperature rose every other day, and at about its seventh recurrence it climbed so high that his body was blazing hot, and in his delirium he muttered words that made no sense to me but which sounded like alternating cries for mercy and boyish high braggadocio. It was almost impossible to get him to swallow the cooling decoction of leaf juices that I had prepared for him. And as I heaped on him the cool wet fig leaves I deplored my helplessness.

At the end of the third week he began to recover; but now he was yellow and miserably thin. It was many days after the fever had broken that a flicker of light came back to his eyes, and for months his face remained drawn and pale. However, Rumsu seemed to emerge from his illness into a phase of greater self-assurance. His eyes no longer turned away shyly or dropped respectfully when he looked at me. His look was one of spontaneous affection and trust, and soon he was scouring the hillside for the roots and tubers which he thought I particularly liked. He also added to our diet out of his own knowledge, choosing the curled fronds of delicate ferns and soaking them overnight for us to eat the next day.

Our mutual affection and his self-assurance encouraged him to manifestations of mischievous playfulness. Often after my meditations or recitations I would realize that he was sitting at a distance of ten or fifteen yards, aping my posture, his face turned in my direction and brimming with quizzical amusement as if asking what on earth could be the meaning or purpose of my occupation.

One morning, after bathing in the stream, I ascended to the bank where I had placed my loincloth and worn shawl under a large pebble. Neither garment was there. I searched for a few minutes till I heard Rumsu burst into laughter on the hillside twenty yards above me. There was no more than fun in his mind, and seeing I had spotted him he immediately brought me my clothes. Another time, while I bathed he quickly washed my garments and laid them out to dry where I had originally left them. But the incident which amused and perplexed me most was one which he repeated two or three times. He would take my clothes up into a tree, and on seeing me searching for them he would sing snatches of romantic-sounding songs in the hill patois. After he had returned the clothes I smiled at him and called him 'Krishna'. He looked at me, his eyes blinking blankly. Had he not heard of what Krishna had done with the clothes of the gopis while they bathed in the river? He shook his head and seemed genuinely ignorant of the story. In a way, too, this hiding of the clothes was, I suppose, Rumsu's loveplay with me.

We must have spent two months under our rock ledge before I felt that Rumsu was strong enough to move up the valley. That raised the problem of his future. His Kangra home must have been at least a hundred and fifty miles over the hills, and I could not possibly suggest that he should try to find his way back alone. Besides, at this time the shepherds were all in the high pastures to the north of the Kulu valley, and not in Kangra.

'What would you like to do now, Rumsu?' I asked.

It was as though he was prepared for the question. He did not have to think even for a moment. Immediately he rejoined in a voice now turning husky, 'What are *you* going to do, my Sadhu?'

'Your Sadhu will simply go where he thinks he should. Probably he will go up the valley of the Beas,' I replied somewhat noncommittally.

Rumsu gave his slender shoulders a quick shrug and said, unconsciously mimicking the tone and manner of my last statement, 'In that case, your boy Rumsu will also probably go up the valley. It suits him well.'

'It suits him well?' I asked, smilingly.

'Yes, my Sadhu. Don't you know that my people have gone up this valley to the high mountains? I will be going towards them,' he replied brightly, so brightly that I decided anything I might say would disappoint him. He sounded as if he expected me to lead him to them.

Later I told him gently that I would probably not go into the very high mountains, remaining near the head of the valley. Wouldn't he therefore like to take up a position in a village near our rock ledge, thus passing the remaining half of the summer, and then with the help of the villagers put himself in touch again with his people on their return journey to Kangra?

But he shook his head vigorously and again replied firmly, as though he had it all worked out, 'My Sadhu has not been in the valley before. I have. At the head of the valley, where we will go together, the gorge is narrow, and I will certainly meet my people when they come down from the pass. Besides, I want to stay with my Sadhu. I like the way he talks to me, and I like his strange food and his singing.'

I smiled and put my hand on his shoulder. Rumsu looked up at me shyly. As I pressed his shoulder he looked up again, and this time his eyes proclaimed that he knew he was victorious. He darted away and danced around in a mad caper. I was nervous for his leg, but the young limb had healed perfectly and responded fluently to his high and noisy spirits.

It was entirely unusual for a Sadhu to be trailing a lay person along with him on his wanderings. But Rumsu was a mere boy and the circumstances were unusual. I could not throw him off and risk his being lost again.

Slowly, through the summer, we walked northward up the gay easygoing valley of Kulu. Rumsu was so clever at locating the forest foods that we kept away from the hamlets. Besides, neither of us wanted other company. Without realizing it I had become very attached to the boy. He had learned another prank which pleased him immensely and endeared him to me. One morning after I had finished my intoning of the Upanishads, Rumsu sat down crosslegged, and closing his eyes — as I did when I recited the scriptures — he intoned, copying my manner, the whole of the Mandukya Upanishad and, except for one or two sounds which probably I myself had acquired the habit of blurring, his performance was perfect. After the recitation his face was beaming, and nodding

213

repeatedly he said, 'The young Sadhu is not bad, don't you think?' Then he burst out laughing, pleased at my amazed smile.

Sometimes, for variety, he would climb a tree and intone snatches from the Bhagavadagita or the Ramayana to the tune of shepherd songs. He would follow this up with shrill shepherds' whistles and call to me, 'My Sadhu, even my sheep will become your followers when I sing these songs to them!'

One day I tried explaining to him the meaning of some verses he had been singing. But it was no use. The thought was dull in his ears. He tossed his head and said, 'Don't bother yourself, Sadhu. The sound is good enough for me. You do the thinking for both of us!'

Passing Manali, the northernmost town in the valley, we found an unoccupied room near some hot springs below the pass. Rumsu said he was sure his people would return by the path at our doorstep. Besides, it suited me perfectly, being a most beautiful setting for quiet meditation and peaceful contemplation. Opposite on the blue-green hills, the stately cedars, singled out by the poet even in the ancient epic Ramayana, stood in dense clumps. Above them the soft green pastures were lustrous in the sunlight which shimmered also on the blue pines on the topmost slopes. To the north the high mountains gleamed with their display of blue-white glaciers. And below us, in the valley, flowed the river Beas, light-flecked like the moving silvery wings of a great dragonfly.

After a few days in this setting my meditation took me into a new realm of experience. A period of absolute silence and stillness would give way to a sequence of beautiful sounds. On the first day there was no indication of the source of this music. On the next day there appeared before me, in my meditation, three great trees. A gentle breeze shook the leaves and I heard a beautiful harmony of sounds —

it came from the falling leaves as they circled to the earth. When they touched the ground the last notes were resonant and deep. I opened my eyes: before me were the first autumn leaves falling to the ground, and as they fell they were playing their beautiful symphony. On the immediately succeeding days my ears were filled with yet more music. There was the orchestra of sound — which normally we hear faintly — produced by the rise and fall of the wind in the pines, over the grass, on the still waters, and swift-flowing rivers. To have the fullness of all these sounds released to my hearing was itself magnificent, but when the full orchestra played to me — the whirring of the wings of birds, the calls of insects and the smallest animals, the gentle little explosions of the soil and the jingling of pebbles under the water, together with the full range of wind sounds — I knew, in small part, what the music of the universe meant.

When I emerged from a sitting of this kind I could feel the immense pleasure within me break into a smile on my face and light it up with gladness. Several times Rumsu said to me, 'The Sadhu is very pleased with life. I wonder what could have happened!'

I would put my arms around the boy, wishing I could open my mouth to pour from it some of the wonders of the music I had heard; but all I could do was to press him dumbly to me. After these new experiences had gone on for perhaps two weeks, autumn was setting in and the flocks were beginning to descend from the passes. At that time there occurred the strangest experience of my relationship with Rumsu.

One evening I emerged rather late from a period of meditation. The sun had set long since, and at that hour Rumsu was always asleep. But that evening he was sitting upon his pallet of straw. I could just see his outline in the dim red glow of our little fire. In a singsong voice he said to me, 'Soon after the Sadhu sat there before his fire, he arose quietly and took me out with him. We went up the hill. Then we stopped, and the Sadhu lifted his hand. Immediately many sounds came from all over the hillside — birds, frogs, stones, grass, drops of water, leaves falling. Then the Sadhu dropped his hand, and the sounds died again. Now I know why the Sadhu smiles so happily after he sits there so still. He pretends to sit there, but he goes away and listens to the sounds of the hills. Then he returns, full of happiness!'

His words so closely described what had occurred during the journey I had just completed in my period of meditation that I was taken aback. I concealed my amazement and asked, 'But how did the Sadhu bring you back, Rumsu?' — wanting to discover everything he could tell me about this journey on which he had apparently accompanied me.

'Does the Sadhu think I have so quickly forgotten?' he replied quickly, with the ingenuous smile of a bright boy who was sure he had the right answer. 'He simply took me by the hand and led me here. Then I must have fallen asleep. But I awoke because I wanted to talk about our journey.'

How could I tell him that he was imagining things, and that neither of us had moved from the room? But how had Rumsu got himself into the adventure with me? I could not explain it to myself. I felt he had really accompanied me, but as to how that had been accomplished — even my own journeys in meditation were a mystery.

On some days Rumsu would ask — after my meditation — why I had not taken him with me. Smilingly I would say I had made so long a journey that it would have tired him. But he would upbraid me with an outpouring of his affection.

'So you would leave me alone, and travel far away? Perhaps you would never come back again. Then what would Rumsu do?' The child was wiser than I, in a sense. It was no good trying to put him off jocularly. It was happier for both of us when he could clap his hands and say, 'That was a wonderful journey, Sadhu!' Then he would recount to me, faithfully and vividly, some part of my own experience in meditation that day.

It was not only during my hours of meditation that Rumsu and I entered into each other's lives. The whole day he watched me closely and with affection. The only time he let me out of his sight was when he searched the hills for our strange victuals. These expeditions he conducted with amazing swiftness, for within twenty minutes or half an hour he would return laden with the best that the hill could provide, his face beaming with joy. The rest of the day he played outside the hut with pebbles, twigs, or whatever caught his fancy. But his eyes were always on me, and every now and again he would rush up to say that the Sadhu made him very happy. In truth, he too made me very happy, and I found that I followed his movements as steadfastly as he did mine. I no longer pondered as to whether it was right for a hermit to delight in the company of a boy. I brushed aside my doubts, telling myself I had had to give him shelter, and I raised no questions about my deep affection for him.

Each day now the flocks were being driven down the passes, and much of this time Rumsu would spend hopping from one high boulder to another in front of our hut, peering into the distance for his people. After a week or more of this peak period of returning traffic from the mountains I began to fear for the boy. Perhaps he was going to be disappointed. But he was confident that his people would soon pass by.

He was right. One day, in the early afternoon, he ran

into the hut from one of his perching places. 'Sadhu! Sadhu! They are coming!' and he tugged at my hand. I got up and followed him to a bluff near our hut. He pointed to a flock about half a mile up the valley. 'There they are, Sadhu! See, that is Raghu — right in front. And at the back is Luchu, who is in charge of the flock. And the dogs, Moti and Joti, too! They are coming, they are coming!'

They were too far away for anyone to identify the men, who were no more than moving specks in front of and behind the undulating grey mass of sheep approaching us over the hills. Again I was fearful for Rumsu as he danced around and flung his arms in the air. A few moments later he was sending out the shrill piercing whistle of the shepherds and calling to them.

'Luchoo-oo-oo-ooh! Raghoo-oo-oo-oo-ooh!'

He waited, and tried again. There was a long moment of dead silence. Then the answering whistle came. Rumsu was wild with joy, but still I could not be certain that he was right.

In ten minutes my suspense was over. The flock approached, and Rumsu dashed towards it, shouting to the dogs and to the men. Soon they were all embracing each other. Dancing by their side he brought them towards me. Then he bounded over to me and tugged at my hand, turning and grinning at his friends. I could see I was on show — a sort of magician who had produced him, and now he was producing me. Apart from his shouting it was a silent drama. His companions were men of few words. They raised their hands in a *namaskar* out of respect for a Sadhu, and they seemed dumbfounded at the reappearance of Rumsu.

Then the boy started tugging them away. He was keen to be on the sheep trail with them again. Nodding his head he called to me as he was pulling them away, 'The good Sadhu has given Rumsu back to Luchu and Raghu, and to Moti and Joti. Thank you, Sadhu, thank you! And make happy journeys over the hills! I wish I could come with you, but I have to go now. Namaskar, Sadhu! Namaskar!'

I waved to them as they went round the bend, just below my hut. That was the last I saw of Rumsu. I knew the day would come when he would have to leave, for if we did not meet his flock I would have to get him back to Kangra somehow. And yet, now that he had left, when I re-entered the hut it was blank and dark, and filled me only with a nauseating sadness. My knees collapsed under me and I lay dazed till the next morning.

From then on in my meditations I would still float away over the hills, but I was no longer alone. Rumsu would be with me, grinning, capering, tugging at my hand, and humming the song of the hills. Since the hut felt so empty, I took to sitting outside during the hours of the day when the sun shone. But my eyes would follow the sheep trail down the valley, and more vivid than ever became the recollection of Rumsu and the painful scene of his departure. Then it became difficult for me to meditate.

I was afraid for myself, and decided that in order to help me tear the yearning memory of Rumsu from my being I should move from our last abode together. It was best to act immediately. I set out, climbing the hillside behind the hut, determined to face the severest conditions that I could stand. In two days I climbed to the region of the mountain passes in the area — above the snow level. my feet frozen and stinging with pain I walked on the ice and rocks, returning to the grass level only for the night, or perhaps finding a stray tree or some boulders to shelter by. Pain, physical exhaustion, and cold numbed my senses and I would fall into a kind of stupor each evening.

As I trudged painfully south-eastward my determination

to re-establish a proper detachment appeared to succeed. After about two weeks of this agony I felt I was in full command of my thoughts again, and I descended into the valley of the river Parvati, near its source. I refreshed myself in the hot springs a few miles lower down, and turned on to the southerly slope of a smaller valley near one of the Parvati's tributary streams. I came upon a spacious rock shelter above the clear sunny waters of the stream. Here, I felt, till the skies became wintry in the middle of December, I could spend some weeks in the lonely beauty of these mountains, responding only to my inner urges for meditation and the recitation and contemplation of the Books.

I discovered chestnuts, ferns, and roots, of which I laid in a stock; and on visiting the neighbouring villages I received a quantity of buckwheat which I also stored. I thus prepared for two months of independence and unimpeded concentration on my life as a hermit.

With all in good order around me, I launched out enthusiastically on the life I had planned. I remember how I recited verses, intoning them strongly, for half a day at a time. My meditations, too, stretched over long spans of hours. This new routine invigorated me, and in the middle of the day I would splash happily in a warm sunny pool in the stream below my sheltering rocks.

It was not I alone that delighted in the warmth of the clear water. By my side was another pool, smaller than mine, and shallow.

Each day on its rim perched a thrush, gleaming black and deep burning gold, on its prim black legs. As I entered my pool and splashed in its waters the thrush would hop into its little pool and thresh the waters with rapid thrusts of its wings and quick excited movements of its head.

At first I was very remote with the little bird, as I

thought that any sign of forwardness on my part would frighten away my delightful bathing companion. But one day the bird's movements were more rapid than ever, and its head bobbed in and out of the water in an amusing jerky rhythm. I burst out laughing at the pranks of my companion. The thrush stopped and looked at me questioningly. Then it decided to splash the water still more energetically. A few moments later it stopped again to look at me. This time it nodded its head and chirruped joyfully. Then it danced through the water again. When I came out of the pool and stood on a rock, letting the water drip off me, the thrush cocked its head and blinked its eyes at me. What was I to do? I could think of nothing except to stretch out my hand towards it. Arching its head it inspected my hand, and decided to reject the offer; but it remained perched on its rock as long as I stayed on mine.

The next day the thrush was there again and nodded its head swiftly as I smiled at it. Again I laughed at its antics in the water. I was in a gay mood. When I finished my own splashing, standing on my rock, I began to intone verses from the Upanishads. My voice responded to my mood, becoming as resonant and musical as it was capable of being. I noticed the thrush studying me with deep attention while I intoned. I remember I stopped reciting, arrested by the sight of an orange and gold horse-chestnut tree higher up in the valley, of a sudden lit up by the rays of the late afternoon sun. The leaves trembled slowly, heavy with the largesse of colour and light. It was at once a vision of bounty and a premonition of the imminent shedding of the leaves which would leave the tree shorn and naked.

I was reflecting on the sudden change which would transform the tree from one extreme to the other, when I heard a deep warble from over my shoulder. I turned around. The thrush had moved to a nearer rock above me, and facing me it was pouring out its song. It was a flowing, sad song, but before closing it rose to a twittering searching for some new mood of joyousness. Then the bird stopped, and jerking its little head towards me in a series of quick movements it seemed to ask what I thought of its music. I nodded back, and again the little bird warbled. In an even lower key now it called for some reassurance of the return of spring, perhaps. Then we both stood silently on our perches, looking at each other for a while before the cold air forced me to seek my place of shelter. As I left the thrush flew up the valley.

That evening I seemed to float quickly into my mood of meditation. I was vaguely aware of gleaming black, pellucid, and burnt gold spaces through which I soared with tremendous swiftness. The speed became faster and faster, till I felt I was a transparent form shooting through the coloured spaces and becoming part of the golds and gleaming black. I came to from my trance with a joyous tingle in my body, and softly I laughed to myself.

The next day the bird had already taken up its position on the nearer rock, and it twittered at me as soon as I arrived. Together we plunged into the water, this time the little creature finding a shallow edge of my pool in which to splash with me. It flailed its wings hard against the water, throwing little jets of spray towards me, and then it jumped to the neighbouring ledge and chirruped delightfully. I laughed happily in response, and splashed my fingers in the water in the direction of the thrush. It entered the pool again, and coming nearer it flailed the water at me with even greater zest.

Bubbling with laughter I scrambled out of the pool to dry myself. I recited again from the Upanishads, and realized that I was lilting the verses. As soon as I stopped the thrush took over, and its warbling notes ended in an upsweeping trill as if in response to my lilting recitation.

I laughed gaily at the end of the song, and the bird hopped responsively on the edge of a rock not three feet above me.

I had begun to descend from the rock to reach my loin-cloth and shawl, when I heard a quick feathery sound above me. Before I could turn around to look I felt the delicate prick of claws on my shoulder. The thrush had found a new perching place. I turned to look at it. The bird bent its head towards my face and nuzzled against the tender skin of my neck under my beard. For a few moments I felt it snuggle there, then it hopped farther back on my shoulder, and in another moment was again on its rock.

The next day, in the early afternoon, the thrush was outside my rock shelter, nodding its head in greeting, and its pulsating eyes beamed their trust in me. I brought out a few grains of buckwheat and held them in my open hand. The thrush hopped on my wrist and delicately picked the grain off my palm. Then it flew up over the rock and I thought it had gone for the day, but a few minutes later it returned to twitter and chirrup at my non-existent threshold, while it spread and gently flapped its shimmering wings. It was time for us to go down to the water, it was saying; and the bird was right, for the days were shortening and the sun would soon heave over the hillside.

Each day the thrush came earlier to my abode, and we ate, sang, splashed, and played together. With our eyes and movements we conversed, and altogether we came to have a life of our own. Before it left in the late afternoon it would perch on my shoulder, and sometimes on my head. When I set aside my daily ration there were a few extra pinches for the thrush, for we were boarding together. And without realizing fully what I was doing I shifted my hours of meditation so as to leave the whole afternoon free for my thrush. Sometimes it would come at an unexpected hour and sit quietly while I meditated. I would open my eyes to find its puzzled but patient open-eyed stare directed at me. Then I would smile and it would burst into a jolly chirrup, as much as to say it was time now to break our mutual silence and be living more fully.

Several weeks passed. One night it clouded over suddenly, and a raw wind, with gusts of cold rain, blew down the valley. In the early afternoon of the next day a sad thrush appeared at my rock shelter. Its feathers were ruffled and the gleam had gone from their colours. It chirruped complainingly, and its eyelids kept drooping over its lack-lustre eyes. I tried to soothe its feelings with sympathetically spoken words, and stretched my hand to it. The thrush came slowly to me and climbed to my shoulder. It nuzzled its head against my neck and pushed its way farther under my beard. Instead of its usual warm palpitating, I felt the cold ruffle of its feathers as it sought warmth.

I could not even offer it the comfort of a fire, for I had no means of lighting one. I took off my shawl and made a cosy nest, in which I left it while I went up the hillside as quickly as I could and collected an armful of leaves. I put two small pieces of rock against the main rock of my shelter and made a little manger which I lined with leaves and grass. There I put the bird. Its eyes brightened a little, and it warbled for a while contentedly. Then it tucked its beak under a wing and lay quietly.

My whole body was trembling with apprehension for the little creature. Why had this illness stricken the beautiful singing, playful thrush? Then it occurred to me that for some weeks I had not seen any similar birds in the valley. The others must have long since flown over the southern hills to the lower and warmer valleys. Then why had this thrush stayed?

I recollected what had happened in the past weeks. I

224

had played with the bird and it had responded, perhaps noticing that I needed company, for I was alone. While its fellows had flown southward it had stayed to cheer my solitude. In our friendship and the sunshine it had braved the changing season, but then the cold and wet of the winter had crashed into our valley and had pierced the bird's minute body. I took some of the dry straw from my own pallet and supplemented the warm leaves in the manger. To this attention the little thrush responded with a faint twitter of thanks.

That night a fine sleet fell and it became piercingly cold. My pallet under me felt like a slab of ice. I lay as close to the bird's nest as possible, hoping there was still some warmth in my body that might be transmitted to it.

All I could do was of no avail. The next morning the bird was dead. My eyes filled with tears, and heavy sobs choked my throat and convulsed my body. I could not stop myself. I felt my own heart had been killed within me. My efforts to allay my sorrow made no headway, and by evening a numbness had settled on my mind and body. It was as though my own life was involuntarily ebbing away to wherever it was that the tender life of the thrush had gone. I could not muster up even enough strength to tell myself that this was a state completely incompatible with my existence as a hermit. I was not even ashamed of myself, and already there was a vague realization that I had to admit that I was no longer in the detached state that was expected of me as a Sadhu. That evening I was unable to meditate. My sorrow held me the whole night long, and my tired body tossed between sleeping and waking.

The next day I sat silently through the morning, trying to compose myself. It became clear that I could not again set out to do as I had done when Rumsu left: I could not clamber back to a true hermitage. I was physically too weak and felt too remote from that state. Besides, what would be achieved if I were to make the effort? Perhaps for a few weeks I would again attain detachment, but a boy and a bird had easily stormed my citadel of solitude. I was much too vulnerable to try again.

Meditation gave place to contemplation, and after five days I was forced to concede that I would have to abandon the saffron robes of the Sadhu.

It was not an easy decision. Those years had produced their own intoxicant and had given me vitality and a wealth of inner experience. Yet the vital chord had snapped. I was still held, but not strongly enough to revoke the counteracting force of the urge of my nature for affection and close companionship. In a daze I sat still under my rock shelter, not moving for two days. On the third day, barely conscious of what I was doing, I set out down the valley. Four days of steady walking brought me into the warmer valley of the Beas. There I hoped I could get a lift on a passing bus; but the first winter rains had so badly damaged the roads that no buses were running. Weary and cold, I struggled on towards Mandi. About a week later I must have been fifteen miles from the town when a bus stopped and took me on. Hitch-hiking along the way, ten days later I arrived in Amritsar, almost seven years after I had left my native city for the Sadhu's grove.

Part VI

I WALKED to my old home. In the front courtyard were a number of my servants and their families. No one recognized the tall, emaciated, grey-bearded Sadhu. Two of the women went indoors and returned with bowls full of rice and wheat flour, thinking I had come to ask for alms. They were surprised that I ignored their offers and walked through the courtyard towards the house.

It was early in the evening, about an hour before the evening meal, and I reckoned that whoever was living in the house would probably have returned from the day's work. Seeing me, one of the servants — whom I remembered as a young man helping about the house when I had left — ran up and called, 'Sadhuji, what is your pleasure that takes you to the doorstep of the house?'

I was startled by the question. It had not occurred to me that I would be questioned as I approached my own door. The servant saw I was taken aback and his suspicions were aroused. Perhaps I was a housebreaker or worse, masquerading as a Sadhu. I had to say something which would immediately check the tide of apprehension.

'Is Narinjan Dasji in?' I asked, certain that he would know who I meant, even if Narinjan was no longer living at the house.

It was now the servant who looked surprised. For a moment he stood transfixed; then his expression softened and he apparently decided that perhaps, after all, Narinjan had asked the Sadhu to the house. He began to move towards the veranda, but suddenly he turned around, the expression on his face having sunk to a kind of abject fear. He bent low before me, and raising his hands in a namaskar he said, 'Come and wait on the veranda, Sadhuji, while I tell Narinjan Dasji that you are here.' Then he went into the house.

Some few minutes later a slightly smaller and bent version of the Narinjan Das I had known, with a face of painted pink and ivory parchment and a startlingly white moustache, appeared in the doorway. The gentleness of his expression had become more pronounced with the passing years. I was deeply stirred by this first sight of him after seven years, and my body trembled with emotion.

His smallish eyes blinked on seeing me, and his lips were forming a question. I got up from the chair on which I was seated, and, smiling, I went up to him. Just before the question on his lips was uttered, and to the accelerated blinking of his baffled eyes, I said, 'Narinjan Das!' and I put out my hands to him.

His jaw dropped, and the parchment face crinkled with animation. Before I knew what was happening he was bending down to touch my feet, as he had done the very first time I spoke to him almost thirty-five years earlier.

'Sir ... sir,' he was saying, fumbling at my knees. 'My Gyan Chand! My God has brought you back!' The last words came in little sobs. I had bent down to raise him, and as I drew myself up I caught sight of the amazed servant running to the courtyard.

Narinjan was sobbing like a child, his head buried in his hands. I put my arm around him. He wept more freely now, as if my gesture had given him express permission to do so. By now the servants had gathered round the veranda and were looking on in bewilderment. Two or three of the eldest were smiling as they came towards me. After a few minutes Narinjan was quiet. I turned to the servants and greeted them. They had taken off their caps and turbans, and their expressions were those of men witnessing an inspiring, almost supernatural, scene. I felt the best way for them to become reacquainted with me was to behave as normally as possible. I had to set about doing something normal, otherwise standing there the deep stirring of my own emotions would have become obvious to them.

I said to Narinjan, 'Come. Let us go in and talk.'

He was still crouching in the position in which he had been crying, but he responded to the pressure of my arm. We entered the house and went into one of the sitting-rooms. Immediately I felt overcome by nausea. My legs were shaky and I seemed to have no sense of direction. With a great effort I steadied myself and sat down on the nearest couch. Narinjan took the chair by my side.

The room was large and the ceiling high, but I was near suffocation. I rested back against the couch and closed my eyes. A thin wisp of energy stirred in me, and I was able to say quietly, 'Narinjan, please get me a cup of tea.'

A few minutes later I was sipping hot tea. Reviving a little, I smiled at Narinjan and said, 'This is the first hot food that has passed my lips in seven years!'

Narinjan looked at me as if in veneration, and his eyes filled with joyful tears again. Hastily I thought of some impersonal subject to bypass the quickly roused emotions of my friend.

'Has all gone well with your business affairs, Narinjan?' I asked.

'Great sir, I have never treated any affairs as mine. All these years I have merely been your custodian, and because they are your affairs they have been blessed by the Almighty. I have done nothing at all.'

I smiled at him. 'Then tell me what you *have* done, Narinjan. I know your words can only mean that you have done a great deal.'

'Yes sir. We have bought two new lumber yards, the mill has been expanded, and it has doubled its production. Anticipating your approval we installed a plywood plant three years ago, and now we are well established in this new venture. You do approve, sir?' he asked.

As he talked I observed him. Though a masklike parchment still covered his face it was the Narinjan Das I knew so well. He was talking about mundane business affairs with the warmth and reverence he had always had for these matters. I realized that that was why he did not need to talk much about himself or of his intimate feelings for other people. His emotions were lodged in the timber yards: they were his life. People he normally ignored, except in naive affection or mistrust or loyalty. He was completely unequipped to face a personal situation such as my return, or the elopement of his daughter Pushpa.

Narinjan, with all seriousness, was now asking for my approval for a factory that had been in existence for three years. It made no sense objectively, and yet to poor Narinjan it was a vital issue.

I smiled and said, 'Yes, my dear Narinjan, of course I approve.'

But I could not share his warm emotion about our business affairs. I led him on with other questions and learned that my elder son, Man Mohan, was well and prosperous. He had built himself a new house in town and his three children were growing up. Romesh, the younger boy, already a prosperous lawyer in Delhi, was now a young Member of Parliament: he had become prominent in the 1942 phase of the movement for independence, and when the country regained its independence he was selected as a Congress Party candidate and had been elected easily. I had been eager to inquire after my sons, and this news filled me with longing to see them again. I asked Narinjan no more questions, and again struggled in silence to gain control of my emotions.

Later, when we got up to look at the house, I managed to provide myself with considerable amusement. I walked through the rooms in little jerky spasms, trying to avoid the furniture, the corners of the rooms, even the walls. Everything seemed to be in the way. It was as if the house were a ship pitching and tossing in heavy seas.

The old mansion boxed me in and was a very poor substitute for my rock shelters or the single room in the grove or on the hillside in Kulu. I felt I could never again accustom myself to being cooped up, much less enjoy it.

My discomfort was deeper than physical. It was not to the house that I had returned. How could I? It had been in me, and I had rejected it. Now I could not give myself back to it. I fled from my bedroom, choosing a top floor room from which I could easily gain the open terrace, and I decided I would use one of the surmounting hexagonal corner rooms as a sitting-room. In my study the books gave the impression of having been swallowed up by the walls. I rescued a few from this appalling fate and took them to the hexagonal room, where the only furnishings were a mattress and a couple of bolsters placed on the floor.

It was not till the next morning, when I was descending from my upper rooms, that I remembered the secret room beyond my study. I hesitated a moment, troubled by the memory of this room. Then I pushed open the door and entered. The servants had already opened the wide French windows to the western balcony, so that in the cool dim morning light the room fluttered on its wings of colour and delicate textures. In the house it was like a rare gem tucked away in one corner of a heap of bric-à-brac. I realized afresh the meaning of art. Here was what man could do. It almost justified the pretentious mansions in which he hid his genuine creations of beauty. I stood on the delicately worked balcony and looked at the morning sun streaming upon the garden through the soft haze of light silvery mist. I turned to the room again and went up to the pictures — they were as beautiful as a galaxy of planets in the dead of night. They were not just small spaces of colour in their narrow frames, but worlds of delight bursting from the mind of the artists. I gasped in surprise: in miniature they were reminiscent of the greatest of my own flights into space in my periods of meditation.

Ignoring the rest of the house I went out to the garden. The firm feel of the earth under my feet gave me a sense of belonging. The sun poured into my eyes, and though I was clad now in my usual clothes I could feel my body assert the habit of the life I had just forsaken and drink in its rays of warmth and light. While my eyes and body were being filled thus, the shrubs and flowers seemed to magnify in size and intensity of colour in my mind. I felt I was being filled with the sensuous beauty of life.

The Persian wheel, in the corner of the garden, stopped — cutting off the noise of its creaking wooden gear and the splashing of the water. The sound of bird-calls immediately filled the air: the garrulous chatter of the maynahs, the greedy chirruping of the hedge sparrows, the raucous crows and jays, an occasional eerie whistle of an eagle, and in the midst of it all the song of the bulbul — full-throated and salty sweet, bringing to the garden the taste of its two favourite haunts, sparkling cool springs and arid brown deserts. I stood listening, and watched these little pulsating creatures darting from branch to branch, along the hedges, and on to the lawn. And they seemed to fly into me. I closed my eyes. Within me was a world of images of wings and streamers of warbled sound. It was as if I had become an animate crystal prism

232

in which the outer scene was being refracted into its essential forms, colours, and sounds.

I heard slow steps approaching on the gravel. I turned around. It was Narinjan walking towards me, wrapped in a white shawl. He must have been unaware of my intoxication for he plunged straight into his own world. It was as though he was taking up again from a previous conversation in the garden nearly seven years ago.

'Gyan Chandji, you left us only after you had fought off the influence of evil. In these seven years Jupiter has never been in the centre of the firmanent. You knew that would be, and you left. And now you have returned when that ominous planet is due to return to his commanding position.'

I interrupted him. 'My dear Narinjan, I must confess that in these seven years I haven't given a single thought to Jupiter — except that, at times, I must have gazed at the beauty of the planet in the night skies!'

But Narinjan's face became very serious and the parchment on his face unwrinkled a little. 'No, no, great sir. The movements of a wise man are always correct. They are always directed by the benign stars. Only foolish people like me are constantly getting in the way of the evil stars.'

I looked at Narinjan, bent and walking a little unevenly now. He looked so humble. And it was not the humility that comes of a recognition of wide relationships in great perspectives: it was an abjectness. I was sorry for him. With his capacities life need not have been lived under the heavy weights he created for himself. But now it was too late to rescue him.

While Narinjan and I were walking in the garden I heard a car in the front courtyard. Narinjan saw me look over the hedge inquiringly.

'Yes, yes. You could not have remembered that sound

at this hour. It is a new one, not more than a year old. My youngest daughter, Sulochna, is leaving for her clinic. She is a brave girl. She finished her medical studies a year ago.'

I recollected the girl's quiet handsome face as I had known it some seven years earlier when she must have been eighteen or nineteen. I calculated her present age and said, 'I suppose she, too, is married now, and a mother.'

Narinjan shook his head and sighed. The wrinkles on his face became agitated.

'Hai, no! She refused to let me arrange a marriage for her. Yet, as God and you have been so good to us, we lack nothing. Certainly she did not need to work! But what was I to do? She insisted.'

I tried to tell him that her work undoubtedly gave her deep satisfaction; and besides, she was still young and would marry when she felt like it.

'But the strange thing is, and it frightens me, Gyan Chandji, that she refuses those who wish to marry her. Many of the young doctors and lawyers want to make an alliance with her, but she won't hear of it.'

I was thinking of Pushpa by this time. I wondered how it had gone between her and Ranjit. I could not ask Narinjan directly. He had been very much upset over her marriage. So I inquired after the rest of the family. 'I hope the other girls are well, Narinjan?' I asked, managing to sound avuncular.

'Sir, my eldest daughter Indira is the apple of my eye. God has given her four sons. All of them are fine boys, and she works for their happiness.'

I wished Narinjan would look at life somewhat differently. As it was the only qualification he conceded to his eldest daughter was that she had borne four sons. But I was anxious to hear what he would say about Pushpa.

234

He sighed deeply and continued, 'I am unable to follow the stars of my daughter, Pushpa. Once in two years they come here. She looks fashionable and happy. Ranjit, of course, looks pleased with life; but they avoid me. I do not know why it should be, but they have only one child — a daughter. I do not know their stars.'

There was no point in pressing him further. I could not expect Narinjan to tell me much about people. His pathologically sensitive nature was so easily hurt, and consequently he preferred to explain people via their stars. I let him revert to the medium in which his emotions seemed to find satisfaction and his whole personality became effective. He told me of the affairs of the timber yards, the prices of logs, of sawn timber, of plywood, and of the immense prospects for the new factory. I was pleased that these developments had made Narinjan happy.

At the end of his rendering of account he turned to me, his beady eyes very shrunken, and said, 'Sir, you know I am nearing seventy. I had to hold charge of all these material affairs till your return. But now I would like to retire and think only of God — as you wisely have done for these past seven years.'

I looked at Narinjan and nodded. 'Surely you can retire whenever you wish, Narinjan,' I said sympathetically, 'and spend your time as you please. Do not let anything upset you.'

He shook his head. I could see that he was very troubled about something. 'Well, perhaps there is not much time now for anything. Look at my face. It is drying up, and I do not know what could bring it to life again!'

I put my arm around him. 'You have had much to do in the past years, and you are tired. But as soon as you are rested, and spend your time as you please, life will flow through you again. Come, let us go and meet the others.' I took him with me to the front or the house, for there had been the sound of the arrival of cars and tongas.

My old friends, acquaintances, and employees had arrived. They seemed genuinely pleased to see me, but all they could do was to smile and raise their folded hands to me. Karam alone acted differently. He was now overgrown and pot-bellied, and his face had coarsened; but there was still a characteristic animal warmth in his eyes. He embraced me, and his whole body shook as he gurgled with quiet welcoming laughter. 'Slim and straight and strong! Gyan, you look as if you had been rejuvenating yourself — while we have stayed in Amritsar to grow old and fat!'

Sita, his wife, looked on. Time had not given her any cushions. She was gaunt and sharp-eyed. She looked as though she distrusted the stories about my having been a Sadhu. She would have preferred to believe that I had been living a life of indulgence and debauch. But I could see her eyes soften a little as she looked at my emaciated body. I remembered how, earlier that morning, after a barber had cut my hair and trimmed my beard, I had undressed to bathe and had been surprised at the brown slenderness of my body. Only my thighs and chest had depth and strength. My face was long and thin, and my eyes were strikingly bright. They peered back at me from the mirror with an intensely clear light. Recollecting the appearance of my body and face I was self-conscious before the large number of eyes gazing at me. I wanted to reach out to them, but their eyes glittered with curiosity instead of understanding. I felt remote from them. The thrush warbled in my ears, and the memory of the birds in my garden filled my mind. I struggled to control my feelings. It was ridiculous that I should feel closer to the birds than to these people. With an effort I opened my mouth to say something to them. For a moment I thought I was going to break into a birdlike warble, but to my astonished relief I was speaking in more or less comprehensible words.

I knew that for many of them my return was like the return to life of someone they had written off as dead or for ever lost; that they would want to be assured now that this was no trick; that it was, in fact, a wonderful miracle which proved to each one present his own theory of the power of God, or ratified his other beliefs. None of them was in the least interested in what I had felt or experienced during the years of my absence. And none of them imagined that, for me, he might be a strange variety of human being who had never wandered off on a path of his own.

Over their heads I glanced towards the courtyard where as a child I had run up to Moti the elephant, and shouted to him to put down the mahout. That act, and my long lonely journey away from these people of Amritsar, became closely interconnected in my mind.

I talked to them of my joy that the sun had continued to shine on Amritsar. I told them that the more one travelled the greater became the stillness within one — but I could see they were unable to understand the meaning of the words I uttered. The look on their faces was completely unrelated to my thoughts. There was nothing there that I could put side by side with Jennifer, who had always looked as if she were about to take off on an adventure of her own; or even with Ranjit, in whose eyes was always a deep restlessness; or with Pushpa, in whose eyes the light had trembled as if seeking to pierce the opaqueness of life. And there was no look among them to remind me, even remotely, of the bright openness of Ramsu's, or of the tender gleaming quality of the eyes of the thrush.

237

I closed my remarks with the expression of sentiments which they could grasp easily: I was glad to be at Amritsar again, but I was, as they could understand, very tired. It touched them to be called upon thus to express their regard and consideration for me. There was a murmur of affectionate sentiments as they left:

'May God grant you blessings at Amritsar, and through you bless us all!'

'Let us help you in any way we can!'

'You must rest now, great sir.'

'Take care of yourself, friend of the poor!'

When they left I went upstairs and rested. That evening I dined in my own rooms, and after spending a short time on the roof terrace I retired early to bed. This became my daily practice, and for that reason it was not till the week-end that I met Sulochna.

She was already in the dining-room when I entered for our midday meal on Sunday. I was feeling imprisoned in the old house, and must have shown my depression and preoccupation. She raised her hands to greet me, and I saw from the look in her strong, large eyes that she noticed that I was at odds with the world. I could sense her effort to remain calm. I liked this indication of her restrained strength, and as I looked at her I saw that it fitted well with the rather wide cut of her handsome face, her shapely eyebrows, and her calm brow. She held herself, too, in a sure unaggressive way. I liked the way she moved and used her sensitive hands. I wanted to hear her voice, wondering whether it, too, would match the impression she had made on me. And when she answered my query, 'So you are Sulochna?' in a musical contralto, without any inhibiting nervousness, I was deeply pleased.

She said in reply, after giving me a slight smile, 'Yes, Gyan Chandji. I came down a few minutes ago to see whether there was fresh fruit on the table for you and

238

Father. He has told me how much you enjoy fruits and raw vegetables.'

Narinjan Das joined us, and when we sat down to lunch I was delighted at the way Sulochna handled him. He was complaining that he had no appetite. She made no direct effort to get him to eat, saying instead, 'It is strange. Though I went out on two cases early today, I too have little appetite. It must be something in the air. Perhaps it would be better if we were to eat outside in the garden these days.'

A lively discussion ensued, which brought back Narinjan's appetite. So obviously was this so that now he expressed solicitude for her. 'My child, you are young and must eat well. Look at your old father, how he is eating.' So he felt he was setting her an example, not realizing that she had enticed him into it.

But Narinjan looked very tired, and after lunch he excused himself. 'Come, Sulochna. Let us go into the garden and see where we might eat on Sundays,' I said to her. She came with me. As we talked I noticed that she would stop to observe the birds.

'They are much freer than we are, don't you think?' I asked, curious to know what it was about them that arrested her attention.

'Much freer. And, of all things that move, they — and not we — seem to me to be the most fortunate. They are nearer the flowers. It's a better scheme of things than ours,' she responded.

We chose a place where our lunch later could be laid. Then her shapely hands worked precisely as she plucked a bunch of flowers for the house. As we came in she told me she was troubled for her father. She thought his heart was not behaving as it should.

That decided me. I followed up Narinjan's own suggestion, and a manager was found for the yards. He was

239

relieved and pleased, and after two weeks of complete rest it became his practice to come each morning to my hexagonal room for an hour to learn the Upanishads. His health seemed to improve; and during the weekends, when I saw Sulochna, I felt she was reassured about him.

Something else seemed to please her. I did not know what it was, but each time I saw her the bloom of her youth seemed fresher and a certain gentle eagerness possessed her. More and more she was able to impart life and vitality to her father. And either she grew closer to nature, or she felt increasingly that she could talk to me about her early morning walks in the country and how frequently she stole up to the roof terrace at night or in the early morning.

I, too, was happier at Amritsar, and I became aware that it was largely because of Sulochna. In her eyes was the look I missed on my return to the town. And as I got to know her it pleased me that her nature was freedom-loving and open to the varied richness of life, whatever its source. But in spite of the pleasure she gave me I continued to feel a certain malaise at Amritsar, and I had not been there three months when I fell ill with a slow fever and a heaviness in my stomach.

We had a doctor come in, but it was not much help. Then Sulochna decided to take me in hand. She must have given little time to her clinic, for she would sit by me and talk, read, and prepare pomegranate juice for me to drink. My temperature fell, and after two weeks or so the fever broke. She knew I would be restive to get up, so she spent more time with me, thus gently keeping me quiet.

By the time I was ready to leave my bed the warm weather had set in; and seeing that I was unable to come to terms with Amritsar, Sulochna suggested that we go away for the summer. She said that if we went to the foothills it would probably not impose any strain on Narinjan's heart.

We took a place at Solon, below the Simla hills, with a large even stretch of lawn where Narinjan could walk without straining himself. It was a delightful summer. Sulochna and I rambled in the valleys, and listened to the birds and the song of the streams.

There were many younger people at Solon, and sometimes I would suggest to Sulochna that she should make acquaintances among them. At first she would smile and shake her head; but she gave no clue as to why she rejected my suggestion, which seemed so obviously in her interest. After a week or so, when I again mentioned her need for companionship or friends, she turned her face full upon me. Her eyes were very bright as she lifted her chin and said, with all the music of her voice, 'Would you deprive me of the intense happiness of growing up with you, listening to and feeling all the things you speak of?' Then, a few days later, she was grave for a moment before saying, 'It isn't just that I learn things. Did you think that was what I meant?' She waited a moment, and seeing from my expression that I had recollected the conversation to which she was referring, she went on, 'I am intensely happy with you — and with Father. Why should I seek anyone else?' She was gentle and tender with me. I forgot my age — she mothered me so. And it was not old age that kindled her eye or explained the tremulous pressure of her touch when I extended my hand to her on our rambles.

Narinjan seemed much better for the hill air. The parchment on his face peeled off, and often he would say, 'The stars are in balance again. A better balance still is before us — I do not quite know what it means, Gyan Chandji.'

But two days before we were due to leave for Amritsar,

241

Narinjan died in his sleep. It was a blow to us both. For Sulochna it was very difficult indeed. She was tormented by the thought that she had decided the hills would do him no harm. I tried to persuade her that she had been right. It could not have been the hills. I reminded her of how much better he had looked and felt at Solon than at Amritsar. Most of all, in the last few months of his life, Narinjan had shed much of his constant anxiety, and neither Sulochna nor I felt — as I certainly had felt during all the years I had known him — that he had to be treated with the kind of indulgence one extends to a person who had not learned the basic steps of life. It was as though, latterly, he had perceived new heights for living, and in those last few months had run forward to gain them so that he could smile and speak of the stars as being in balance. After several days of heavy silence Sulochna was able to accept this, and her sorrow ceased then to immobilize her.

We left Solon, but neither Sulochna nor I mentioned the future directly. She had her profession, and she spoke of the many things to be done at the clinic. Her work would claim much of her time and energy. For the rest, her character was not that of a dependent person. But we did not explore these matters fully. I think both of us were conscious of a feeling that beyond a certain point plans expressed in words became no more than brave, even heroic, midgets, trying to cover all the spaces of the universe and the even greater voids within us. We looked at each other, knowing that feeling alone could find a way in these immeasurable dimensions.

When we returned to Amritsar, Sulochna — at the pressing suggestion of Pushpa and Ranjit — moved into their town house which they rarely visited. As for myself, I was convinced that I should leave the town. I had returned from my years in England determined to find duties, responsibilities, and a living sense of continuity in the family mansion. I could smile now when I thought back to the ardour of my feelings at the time. But the intervening years had been in some respects — particularly in regard to the goals I had set before myself — a failure. In a sense I had left Amritsar with Jennifer. My marriage had never succeeded in bringing me back to the family mansion; and my seven years at the grove and in the mountains were not an escape but merely an expression of the greater unreality of Amritsar.

But what was I to do with the house? There was no particular reason for selling it. I did not need the money. As Narinjan was no longer there to do it for me I looked into the accounts and other business documents. Two of the yards were his — and now went to Sulochna and her sisters. The other two yards were mine and were doing excellently. Apart from these there was the plywood factory — not counting the paper factory which I had long ago given to my elder son. The income from the plywood factory, my urban property, and my lands was much more than I could possibly want. There were also large sums in reserve in various banks. These amounts had swelled during my years of absence, assisted by Narinjan's careful and scrupulous management of my affairs.

I was somewhat abashed at finding myself in possession of all this wealth. I decided to give the mansion to the Ram Singh College for their post-graduate work; and, to defray the cost of upkeep and provide a substantial contribution towards the expenses of the college, I set aside with the house the income from my two timber yards. In this way I settled the problem of the house and also rid myself of the burden of some of my other properties.

That was about a year ago. Immediately it was all arranged I came away to Delhi and lived in a hotel till I could find a quiet house to my taste. We both liked this

243

unpretentious house, particularly the spacious secluded back garden, and moved in here a month or so before you arrived home from the hospital. A week or so after your return I met you, and since then we have seen each other almost every day, and the story of my life has been lived before you, with you.

Part VII

THAT was Rai Gyan Chand's story. He frequently
referred to it as 'my journey'. Having travelled
with him breathlessly during those days when he
went over it again, I was deeply glad that it had brought
him to the white house opposite my home. Here, in a
physical sense, he appeared to have come to rest. But I
kept feeling that the story was incomplete, and that he
had been reticent about the thoughts and feelings that
had governed its latter phases.

Gyan Chand's purpose in life had not been to come
physically to rest — that was clear. Then what had given
him this sense of finality in the white house? And I was
also popping with curiosity about Solochna, and about
Pushpa and Ranjit. What had happened to them? And
who was the handsome but rather untidy woman at the
white house?

At the end of Gyan Chand's story I felt so close to him
that it seemed natural and proper that I should ask these
questions. And yet I was restrained by the fact that he
had chosen not to clear up these and other matters in
what was, after all, his story. Perhaps he had some un-
disclosed reason to keep certain things to himself.

But my curiosity was stronger than my reason, and I
found myself putting the questions to him. My hesitations
and the delicacy of the situation led me to begin with the
least personal of my questions. I asked whether the move
to Delhi from Amritsar had turned out to be what he had

expected, and whether he felt he had done wisely to leave his old family home.

Gyan Chand's bright greyish eyes looked very remote for a moment, and while he replied occasionally this remoteness reappeared, but was explained by what he said.

'I hesitate to reply too definitely,' he began, 'because we have not been here for any length of time. And it is this that has kept me from saying anything to you about it. However, I do want to try to put into words how I feel about having come to Delhi.'

'Before I came here, in all the things I set out to do in my life I gave myself to them more or less completely. That was true perhaps from my early peremptory interference in the affair of Moti, our elephant. Then in the period of my youth spent in England — I was practically, as you remember, swallowed up for good and all in that experience.

'In the same full way I gave myself to the idea of marriage, and thereafter to the decision to be with the Maid of Amritsar, and to the life of a hermit. But none of those forthright turnings satisfied me. Always too much was missing: indeed, on each occasion, I put away the past and sought from the future all that my new course alone could yield. Up to the time I left Amritsar for the Sadhu's grove no phase of my life completely excluded other impressions, experiences, and intimations. But each did produce an imbalance in my scheme of things.' While he had been saying this the remoteness came back to his eyes. Then he went on.

'Do not imagine that I am evading your question, but I must search in the far recesses of my life for the full answer. I was telling you of the recurring imbalance of my life. But when I went to the Sadhu, and in the years of that whole phase, it was not just an upsetting of the balance. In submitting myself to the discipline of the life of *sanyasashrama* I had to cut myself off totally from other experiences and impressions. I thought I was achieving what I sought and had made my goal throughout life — complete absorption in my chosen occupation. But this attainment turned out to be illusory. Just when I had become a seasoned practitioner in the art of meditation and of the solitary life, I found myself slipping into a deep attachment with the boy Rumsu; and then, when I struggled free, the even more intense and completely unforeseen relationship with the thrush. It was difficult and painful to admit defeat at the age of close on fifty-five. But I realized that to acknowledge the validity of experience was not defeat. So I returned to the world of people and affairs.

'But it could not be a return to the life I had been trying to live — a life of total absorption in one thing or another. I was not absolutely clear about this when I returned to Amritsar, and that was why I suffered from physical and psychological illness. However, gradually it did become clear to me, and then I began to see what I was to do.

'I was to open — to free — all my capacities for life: for affection, thought, work, and understanding. At my age it was difficult to face this conclusion squarely. And besides, I was troubled by the other aspect of the matter: if I dispersed myself thinly over a wide field I could easily become ineffective, and achieve nothing at all. But deep in me these objections seemed flimsy and beside the point. They stemmed from a view with which I had no sympathy — one that stresses achievement, 'achievement at any cost'. To me it seemed that if one could learn to open all one's personality, then one could make available one's full sensitivity to life and not just a sharp probing focus of some of it. That is what I have tried to do. I do not know whether I will achieve anything in the way people measure these matters; but I am wide open now. I love without burdening the beloved with the false attempt to focus my whole life on her. I feel I begin to understand life somewhat.'

He stopped, his eyes lit up, and a smile puckered his face. Then he went on. 'There is something else that I learned. I started out with everything that a man is said to require: health, good looks, intelligence, some natural sensitivity to beauty, and a sufficiency of material goods and money. And yet, whenever I looked at my life, it was almost empty. So I went from one experience to another, always losing out, and pitying myself. It was a lack of courage that afflicted me at many great open gateways to life: Jennifer, Basanti, my studies, Askari, and Pushpa. I felt love coursing through me, and many times in my life I knew of a converging urge — particularly in Pushpa — and yet foolishly I would convince myself of some great obstacle. It was only when I acknowledged to myself the failure of my life as a Sadhu and mustered enough courage to think back that I really entered the first gateway of my adult life. I realized then that my early instinctive bravery with Moti the elephant had been only a childish victory. I had secretly revelled in it, and it had opened to me the doors of childhood and adolescence. But it was useless in adult life, for I was dealing no longer in the realm of elephants but of people: I was slow at realizing this, and instead of entering the gateways at your present age I had to wait until a few years ago.'

He stopped again, his face glowing warmly. Then he raised his eyes and watched the birds in the garden. Leaning back he whistled to the tits that were alighting on a tree in front of us.

Then he went on. 'You see, I am in love with these birds! The thrush opened my bird soul! But seriously, it is so exhilarating to be wide open in all directions. And now in this state of so-called wide dissipation, I work harder than I ever did before. You see me copying the wonderful *Crest Jewel* of Samkara. In the evenings I write a commentary on it. And in my old age I have taken to painting. You asked me once whether I had ever painted, and I shook my head. That was true of the past; but now my mind's eye is so full of the wealth and vividness of colour that I saw — I did not just see it, I lived it — in my meditations, that I try to put some of it down on canvas.

'I do not try to formulate a hierarchy of my capacities and needs. That seems to me to be a fault of the old way in which I lived. There are aspects of life which are especially close to me. The love of man for a woman is obviously of this character. But these intimate things can be so easily spoiled if the rest of one's life goes wrong. Then one focuses heavily and the result is bitter frustration in love. But with my wife I have, I think, a good relationship. It embraces all that I longed for and dreamed of with Jennifer, with my first wife, the Maid of Amritsar — more vaguely, without personification. My other activities and complete sense of freedom of life make possible a wonderful love relationship. Fortunately my wife understands my approach and shares it — I have told her, too, the story of my life.'

I felt he had answered my questions. It thrilled me that at the end, or at what might have been an ending of despair and defeat, on the hillside near the valley of the Parvati, Gyan Chand had had the courage to gather together, as it were, the sum total of his experience and set out again. I could not say, any more than he, whether he was going to achieve something that the world would acknowledge. But that to me, as it did to him, seemed an irrelevant and idle speculation. What I did know was that as I searched among my other acquaintances and friends none of them had either the aliveness or the gentleness of this elderly man. There was an exquisite translucency about his mind, with which flowed also his warmth of feeling so as to create a sense of the unity of life and the tenderness of all its myriad relationships. His eyes sparkled genially, his face was kindly, and his movements had a gracefulness which had been unimpeded by age. Secretly, and with a desire much more intense than that which accompanied my other ambitions and stirrings, I wished that I too would have the courage to live in such a way as to accomplish the fullness which Gyan Chand had achieved.

There remained other questions in my mind. I left them till the day before I was to leave Delhi, hesitating till the last. He had not told me when he had married his present wife and where he had met her. I asked him these questions, stumbling over the words in my own uncertainty as to whether I should have done so. Gyan Chand laughed gently at me, his eyes sparkling, and a deep ruddiness mounting to his cheeks.

'But I assumed you knew! And now I realize that you have been misled because I call her "Devi". She is Sulochna. So I need hardly tell you where and when I met her. Alone, I left Amritsar for Delhi. But my love and regard for her were persistent, and I wrote and told her so. It was a risk but it came off. She replied, affirming that her feelings were at least as strongly involved with me. I suggested that she come to Delhi so that we might talk about our lives. She took a room in the hotel in which I was stopping. Little by little she told me about herself. She wanted to continue in her profession, but she felt it necessary to live so that she could see me and be with me.

'Then I told her my story. It took several weeks, because I did not wish to conceal anything that was relevant to the way my life had developed. I knew she meant it when she said she could not conceive of a way of life more to her liking than that which I now thought I should live. The decision to marry then seemed to follow naturally. I was pleased that it had not been our direct objective — we reached it as a consequence of the fullness of love and the reasonable certainty that we moved in a common rhythm.'

Gyan Chand stopped and smiled. Then he added, 'Of course, I tried to tell myself that I should hesitate because of the difference in our ages. I reminded myself that I had hung back when I had been enamoured of her older sister because I felt then that I was too old. How much more now should this consideration determine my actions. But the argument, in the face of her feelings and mine, was meaningless. It, not I, seemed to belong to the wrong age. I just forgot about it!

'You have not seen much of Sulochna as she has been taking things easy. In a few months' time she is to be a mother. She is, I feel, deeply pleased. And as you might have suspected we brought with us the furnishings of the secret room in my old house at Amritsar. Now they beautify a room to the west, which looks out on a private veranda and then on to a small garden hidden from the rest of the house by two high hedges. And I have been glad of my vast study about love and sex. At last I know the fullness of its application.

'Now you know everything, my dear young friend! But perhaps there is just one other matter I might add. Now, I do not have any regrets in my personal life about the gateways that I did not enter. Frequently you must have seen here a tall, elderly Sikh — he walks with the aid of a stick — and by his side a vivacious beautiful woman. They are Ranjit and Pushpa. Often we visit each other, and the four of us are happy together.'

Then he added with a smile, 'And, undoubtedly, you have noticed the lively beautiful girl of seventeen with them — their only daughter.' I smiled, and Gyan Chand gave me a deep but playful look.

The next day I left Delhi, full of recollections of Gyan Chand, and amazed and pleased at the present of twenty-eight pictures which he had drawn for me. Almost immediately after my arrival at Bombay I began to set down the story he had told me. At the end of a year I had completed it. During that period I had not written to Gyan Chand — partly because I generally tend to be too absorbed in my current pursuits to encourage letter writing; but also because, as I wrote his story, I felt myself to be in very close communication with Gyan Chand and did not feel the need of the additional medium of correspondence. Nor had I heard anything of Gyan Chand during that year. Now I was looking forward to seeing him again, and for a while I wondered whether I should show him what I had written. I decided I should, for our relationship was a frank and open one.

I began to arrange my schedule of work at Bombay so as to leave a free fortnight for a visit to Delhi. It was at that time that one of my mother's letters mentioned Gyan Chand. My eye caught his name in the middle of the page and, eagerly, I turned to the commencement of the paragraph.

It is sad that the white house opposite us should be vacant and locked up again. Your friend, Gyan Chand, died suddenly, and his widow and their infant son have left — no one seems to know where they have gone. It will be such a pity to see the garden neglected and running to seed again.

Gyan Chand dead! I could not believe it. He had looked so well and strong. What could have happened? I wrote at once to my mother and begged her to give me all the facts she could discover about his death. Then I wished I had not written. He was dead, and not knowing how it had happened in time I could formulate some acceptable explanation; or I would just remember him as he had lived, and the recurring memory of his death would be like the inevitable return of the symptoms of a deep-seated, malignant malady — pure, senseless pain.

A few days later, my mother's reply came.

They say that soon after his wife returned home from the hospital with their infant son, Gyan Chand began to spend more time alone — walking in the garden or sitting on the back lawn and working. I remember now that I, too, was surprised to see him looking downcast — so inexplicable a change when his wife had just presented him with a son. Then, if you please, in the middle of the winter he took to sitting out till late into the night, in the back garden — all alone, and often he was heard muttering to himself. They say he aged terribly in three months, and would wander off incoherently in his speech — poor old thing (though he was probably younger than either your father or I). One evening he insisted on staying out all night, and the next morning they found him dead. No one knows why his wife let him do all those mad things! They say she was so absorbed in the child that she forgot all else. Anyway, it is a great pity, and very sad.

So Gyan Chand's journey was over — his journey towards love. Had he reached where he had wished to get? It would seem not. And yet, I had witnessed the fullness of his life at Delhi. Could anyone be expected to live on the crest indefinitely? I wonder what Gyan Chand's answer would have been.